Urban Climate Justice

Urban Climate Justice

THEORY, PRAXIS, RESISTANCE

**EDITED BY
JENNIFER L. RICE, JOSHUA LONG,
AND ANTHONY LEVENDA**

The University of Georgia Press
Athens

Library of Congress Cataloging-in-Publication Data

Names: Rice, Jennifer L., 1981– editor. | Long, Joshua, 1979– editor. |
    Levenda, Anthony Michael, 1987– editor.
Title: Urban climate justice : theory, praxis, resistance / edited by
    Jennifer L. Rice, Joshua Long, Anthony Levenda.
Description: Athens : The University of Georgia Press, 2023. |
    Series: Geographies of justice and social transformation; 57 |
    Includes bibliographical references and index.
Identifiers: LCCN 2022042603 | ISBN 9780820363776 (hardback) |
    ISBN 9780820363769 (paperback) | ISBN 9780820363783 (epub) |
    ISBN 9780820363790 (pdf)
Subjects: LCSH: Urban ecology (Sociology) | Climate justice. |
    Sustainable urban development. | Urban policy. | Climatic changes—
    Government policy.
Classification: LCC HT241 .U6944 2023 | DDC 307.76—dc23/eng/20221201
LC record available at https://lccn.loc.gov/2022042603

CONTENTS

Urban Climate Justice

INTRODUCTION

Realizing the Just City in the Era of Climate Change

The Urban Politics of Climate (In)security and (In)equality

JENNIFER L. RICE, ANTHONY LEVENDA, AND JOSHUA LONG

The Right to the Climate-Just City

In October 2021 thousands of people gathered in the streets of Glasgow, Scotland, during the COP26 conference to demand action on climate change, including many Indigenous activists calling for an end to exploitation of nature by Western corporations and a centering of Indigenous ways of caring for the environment (Lakhani 2021). In October 2019 residents in dozens of African cities, including Cape Town and Nairobi, participated in global CLIMATE strikes, arguing that it was time to address the disproportionate impacts of climate change felt by people in the continent (Dahir 2019). Likewise, the organization Shut Down D.C. closed fifteen intersections in Washington, D.C., in September 2019, focusing on disrupting locations near fossil fuel companies and their lobbyists' offices (Moyer, Tan, and Hedgpeth 2019). These are just some of the hundreds, perhaps thousands, of instances of people taking to city streets around the world during the past several years to demand action on climate change.

In this context, the city has become a key site for the climate justice struggle (Castán Broto 2019; Rice 2016; Romero-Lankao et al. 2018). Yet cities are more than just sites for action: they are also places where uneven vulnerabilities and inequalities related to climate change are produced and reproduced. They are places where access to secure futures and resilient infrastructures are mediated by global flows of capital into and out of cities. They are places where local governments count carbon in ways that have the ability to highlight, or hide, the residents and institutions who are the real source of the climate crisis. As the COVID-19 pandemic—which began during the writing of this book—also shows us: "Both climate change and the coronavirus pandemic have uneven, unequal and long-lasting impacts that depend on where you live, who you are, and what you have" (Sultana 2021b, para 2). These evolving inequities, ongoing injustices, and emerging crises beg a new set of questions about the relationship between climate justice and urban justice.

This edited volume considers who and what makes the just city in the era of climate change. Critical urban scholars have provided robust theoretical and empirical contributions for understanding social (in)justice in cities during the last several decades. At the same time, critical human-environment scholars have examined the uneven socioenvironmental impacts of climate change and associated policies on cities and urban residents. Yet better understanding the intersections of urban justice and climate justice requires more persistent and engaged study.

We and the other authors in this volume are concerned with how we can create urban futures that are secure and equitable for everyone (see figure 0.1, scenario 4). We believe that creating a just city in the era of climate change is a normative political commitment that requires radically different forms of urbanization, development, and governance than are currently dominant in many of the world's cities. Fundamentally, we and the authors who contribute to this volume will argue that climate change should not be understood as a distinct and relatively recent socioecological phenomenon. Rather, we will demonstrate how and why climate change is connected to the long and persistent histories of (settler) colonization, environmental racism, and heteropatriarchy that structure our cities. We and the other contributors to this edited volume also seek out forms of knowledge often characterized as "on the margins" to help us imagine and create alternative urban futures; it is in Indigenous expertise, the experiences of those subject to ongoing environmental racism, and the work of female and queer activists that we can imagine and implement alternative climate futures.

While all of us contributing to this volume are institutionally situated at universities in the Anglosphere, we seek to use this position of privilege to center the voices and experiences of peoples and communities on the front lines of climate change who are both most harmed and most knowledgeable about alternative ways of being. We also seek to use our resources and privileges to challenge the dominant status quo that currently benefits from action (or nonaction) on climate change. We attempt to enact what Sultana (2021a) calls critical climate justice: "dismantling fossil fuel dependency, defetishization of endless capitalist growth on a finite planet, challenging non-participatory democracy, resisting extractive exploitation of natural resources, confronting racial capitalism and Indigenous erasure, among other things." We believe that scholars should theorize these processes but also put forth ideas for real mate-

MORE SECURE

(1) Secure but inequitable *Resilient enclaves for the wealthy* *Carbon / climate gentrification*	**(4) Secure and equitable** *Green New Deal* *Energy democracy and public renewable energy* *Indigenous resistance, Black Lives Matter*
Climate modification and techno-fixes	
Closed borders *Militarized climate hazard responses* *Poor coastal areas under sea level prone to climate hazards*	*Large-scale drought, flood, fire* *Initial impacts of Hurricane Sandy*
(2) Insecure and inequitable	**(3) Insecure for all**

LESS EQUITABLE | MORE EQUITABLE

LESS SECURE

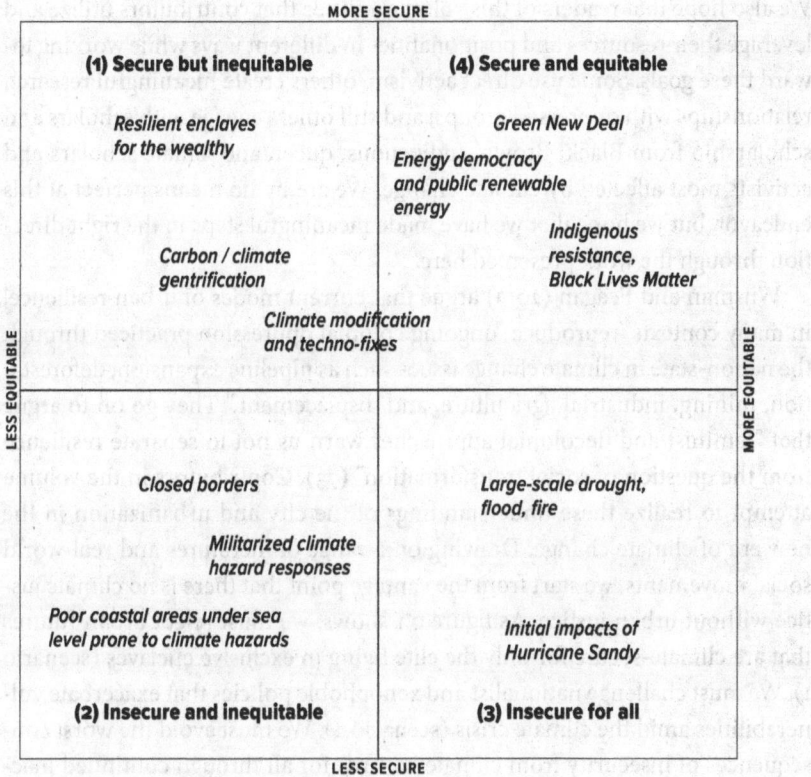

FIGURE 0.1. Possible urban futures under climate change

rial changes that can be implemented across scales and time frames. As we will demonstrate in the coming pages, this should include:

- Abolishing racist policing, carceral, and legal systems that criminalize those made more precarious by climate change.
- Investing in low-carbon social housing and other collective infrastructures that reject private property as the only means to climate mitigation or resilience.
- Paying reparations in the form of direct compensation to and investments in racialized communities and nations who bear the brunt of carbon colonialism.
- Returning land to Indigenous communities and banning western land grabs touted as mitigation or adaptation strategies.
- Centering and caring for BIPOC, female, and queer experiences and knowledges in climate adaptation and environmental restoration.

We also hope that readers of this volume will see that contributors utilize and leverage their resources and positionalities in different ways while working toward these goals. Some use direct activism, others create meaningful research relationships with oppressed groups, and still others engage with scholars and scholarship from Black, Brown, Indigenous, queer, and female scholars and activists most affected by climate change. We are by no means perfect at this endeavor, but we hope that we have made meaningful steps in the right direction through the work presented here.

Wijsman and Feagan (2019) argue that current modes of urban resilience, in many contexts, reproduce "ongoing colonial oppression practiced through the nation-state in climate change issues such as pipeline expansion, deforestation, mining, industrial agriculture, and displacement." They go on to argue that "feminist and decolonial approaches warn us not to separate resilience from the question of social transformation" (73). Contributors in the volume attempt to realize these understandings of the city and urbanization in the new era of climate change. Drawing on a range of literatures and real-world social movements, we start from the vantage point that there is no climate justice without urban justice. As figure 0.1 shows, we must reject urban futures that are climate-secure for only the elite living in exclusive enclaves (scenario 1). We must challenge nationalist and xenophobic policies that exacerbate vulnerabilities amid the climate crisis (scenario 2). We must avoid the worst consequences of insecurity from climate hazards for all through continued inaction and indifference (scenario 3). In other words, we must take seriously the interlocking processes of uneven urban development and climate vulnerability that create a complex patchwork of climate (in)justices within urban areas, as well as between them.

We take direct inspiration from the "right to the city" (RTC) concept, as both an academic theory and an activist movement (Attoh 2011; Harvey 2003). Academics and activists evoke RTC to argue that urban spaces should be recreated to be more inclusive, accessible, secure, and affordable for all people to create genuine forms of community and self-determination. Yet given the urgency and severity of the already occurring climate crisis, we believe it is important to expand this to a right to the climate-just city (Cohen 2018a). That is, we should not only reduce the dramatic forms of discrimination, segregation, and inequality that characterize the neoliberal city but also create cities that will begin to address the climate crisis and protect all residents from climate harm. Importantly, the Right to the City Alliance uses an intersectional approach to urban justice that focuses on not just land reform but also economic, Indigenous, racial, criminal, and environmental justice. Focusing on radical democracy, their mission states that we need "a new kind of urban politics that asserts that everyone, particularly the disenfranchised, not only has a right to the city, but as inhabitants, have a right to shape it, design it, and operationalize an urban human rights agenda" (RTTC 2019).

The RTC concept has been taken up by a number of activist groups who inspire or work directly with many of the authors in this volume. We will engage this project, as it relates to climate change, in this introductory chapter from several standpoints. First, we will review the emerging literature on climate urbanism, which helps us understand the constraints posed by the increasingly close integration of urbanization and neoliberal forms of climate governance (section 2). Then, we examine the relationship between the ongoing processes of colonization, environmental racism, and heteropatriarchy that structure our cities and how this relates to the problem of climate change (section 3). Next, we further describe a typology of four possible (and already occurring) urban climate futures (figure 0.1) that may be experienced under different political, social, and economic arrangements (section 4). Finally, we outline and describe the individual chapter contributions in this book that help us imagine, create, and enact more just urban futures under climate change using theory, praxis, and resistance (section 5).

The Rise of Climate Urbanism

Research at the intersection of cities and climate change is now a well-established field. The 2003 book *Cities and Climate Change: Urban Sustainability and Global Environmental Governance* by Harriet Bulkeley and Michelle Betsill was one of the first publications to critically evaluate the role of cities in multilevel climate governance. Subsequent research demonstrated several important features of urban climate governance, including how local climate governance is deeply connected to the neoliberal state (While, Jonas, and Gibbs 2010; Whitehead 2013); why urban climate governance relies heavily on problematic forms of individual behavior change (Rice 2014; Slocum 2004); what new forms of carbon governmentality and citizenship are created by local emissions accounting (Paterson and Stripple 2010; Rice 2010; Rutland and Aylett 2008); and how cities serve as laboratories for policy experimentation related to climate change and infrastructure (Castán Broto and Bulkeley 2013; Levenda 2019).

Building on this literature, the concept of "climate urbanism" has more recently emerged as a way to understand how and why urban responses to the climate crisis have materialized in a neoliberal, post–financial crisis context, and at a time when the world's cities were already struggling to uphold basic principles of social justice and environmental sustainability (Castán Broto, Robin, and While 2020; Long and Rice 2019). Specifically, the inequalities present at the turn of the twenty-first century—especially urban inequalities— are becoming exacerbated by the shocks and stresses of climate change. Perhaps more worrisome, it is becoming increasingly apparent that attempts to mitigate and adapt to these stresses are introducing new mechanisms that ultimately entrench old injustices (Caprotti 2014; Cohen 2018b). The idea of climate urbanism has come to signify "a policy orientation that (1) promotes cities as the most viable and appropriate sites of climate action and (2) prioritizes efforts to protect the physical and digital infrastructures of urban economies from the hazards associated with climate change" (Long and Rice 2019, 992). In recent years, those efforts have become increasingly framed not as opportunities for just development and equitable forms of adaptation but as investment opportunities, particularly for financiers in the developed North (Long and Rice 2019).

A "win-win" narrative of sustainability and green economic growth has been fundamental to the success of these urban climate programs, as it provided policy makers the means to transform complex environmental and social issues into clear outcomes while also facilitating greater municipal control over politically resistant elements (Long 2016). This narrative also allowed cities to advance new programs and projects that increased city marketability, appealed to capital investment, and provided a desirable set of lifestyle amenities for creative and tech workers who were increasingly becoming valorized as the greatest innovators and economic contributors of the labor force (Rice et al. 2020).

This drive to promote green growth, however, left little room for critical planning perspectives that considered the negative social consequences of these policies. The greening of urban neighborhoods, the attraction of tech-oriented, creative industries and labor, and the implementation of digital platforms and technological modernization ultimately resulted in housing unaffordability, economic polarization, displacement, and environmental gentrification (Dooling 2009; Quastel 2009; Checker 2011). The 2007/2008 financial crisis only further underscored a stark landscape of urban inequality, as elite eco-districts in central smart-growth neighborhoods drew sharp contrast to increasing suburban poverty, housing foreclosures, and poor enclaves with diminished access to services and amenities (Berube and Kneebone 2013; Cooke and Denton 2015; Hochstenbach and Musterd 2018). Furthermore, these post–financial crisis landscapes of injustice and spatial segregation were becoming remarkably visible at a time when climate change—and the need to plan for more frequent and intense climate hazards—was becoming an urgent priority for planners and policy makers.

This global shift has become solidified in recent years, and in its current state facilitates the transition from an era of sustainable urbanism to the current era of climate urbanism in several ways. These include a focus on protecting the status quo of economic productivity in cities against the threats of climate change, a fetishization of carbon as the primary cause of climate change (divorced from the sociopolitical conditions under which it is unevenly emitted), and increased legitimatization of problematic forms of surveillance, segregation, securitization, and regulation of urban residents in the name of climate change (Long and Rice 2019).

Because climate urbanism exists at a time of increasing internal and cross-border displacement throughout the world—which is, in part, a result of the climate crisis—some cities will be forced to negotiate the realities of providing basic needs and services to a large influx of climate-displaced migrants. As this trend continues, it could potentially result in the largest mass migration of humans in history (Rigaud et al. 2018; Podesta 2019), and concerns over this future are giving rise to a global increase in ethnonationalism, anti-immigration sentiment, eco-fascism, and calls for increased border security (Chaturvedi and Doyle 2015; Lawton 2019). The xenophobic trends signal a continuation of the dehumanization and violence associated with racist, heteropatriarchal, and colonial mentalities.

It should be said that this potential for exclusion and segregation represents the current, dominant trajectory of climate urbanism, but as other scholars of climate justice have noted, there are alternative, resistant modes of climate urbanism that also exist. As Robin and Castán Broto (2020) have observed, the dominant narrative of climate urbanism largely reflects a persistent colonial mindset toward urban development and draws heavily on scholars utilizing and reproducing Western perspectives. This fails to acknowledge the multiple ways in which other urban dwellers and scholars—particularly those in the Global South—respond in creative and radical ways to the climate crisis. Instead, Robin and Castán Broto's (2020) call for a transformative climate urbanism that adopts an anti-racist and postcolonial response to the climate crisis mirrors our thinking in this book. Indeed, the consistent disregard of alternative, non-Western, and radical perspectives is a major concern for us, and this book attempts to deconstruct the dominant view of climate urbanism to move us closer to a transformative view of urban climate justice—a project that requires us to engage with a broader set of ideas and knowledge producers, which we turn to next.

Legacies of Colonization, Racism, and Patriarchy for Climate Injustice

From deadly floods in South India, to ongoing severe drought in South Africa, to destructive hurricanes and wildfires in North America, it is increasingly clear that the dangers of the climate crisis are highly uneven both socially and spatially. Climate justice—urban or otherwise—can only be achieved through recognition and repair of the uneven impacts of the climate crisis that are based in raced, classed, and gendered inequalities. For example, it has been documented that 80 percent of people displaced by climate change are women (Halton 2018), while heat-related mortality is higher among elderly and low-income populations (O'Neill et al. 2009). Households living in informal settlements (e.g., slums and favelas) are not only more vulnerable to climate-related hazards but also have lower adaptive capacity because of persistent poverty (Pandey et al. 2018). Two of the most vulnerable cities to climate change, according to a global risk consulting firm, are Lagos, Nigeria, and Manila, Philippines (Law 2019). This is compounded by the fact that efforts at climate change mitigation and adaptation have also been shown to benefit certain populations more than others, even harming some of the most vulnerable urban residents. For example, lower-income and nonwhite populations are disproportionality displaced by ecological gentrification as wealthier and whiter residents relocate to urban centers in an effort to live a "climate-friendly" lifestyle (Rice et al. 2020; Bouzarovski, Frankowski, and Tirado Herrero 2018). Shi et al. (2016) note that adaptation projects often "entrench unequal power distribution by taking advantage of disasters to relocate disadvantaged populations from urban centres or investing scarce public resources in areas of high economic value without giving commensurate attention to historically neglected neighborhoods" (132). Queer and trans people can experience additional harm and violence in disaster response associated with climate change because of heteronormative, or even religious, approaches to assistance (Gorman-Murray et al. 2017).

To better understand these complex patterns of urban inequality, insecurity, and injustice produced under climate change, we argue that we must focus on three interlocking processes—colonization, environmental racism, and heteropatriarchy—that structure this uneven vulnerability to climate change.

What we offer here is only an entry point to these complex and deeply established processes of oppression, but it is important to highlight how they relate to climate change. Ugandan activist Vanessa Nakate offers a crucial perspective here: "There is so much to learn about the climate crisis, and learning about the climate crisis means learning from the voices that are on the front lines. And we have seen how continuously activists from the global south, who are speaking up from the most affected communities—their voices are not being platformed. Their voices are not being amplified. Their stories are being erased. . . . This is a problem. We can't have climate justice if voices from the most affected areas are being left behind" (NPR 2021).

First, we must recognize that the production of urban space is deeply connected to the logic of colonization. That is, colonization and settler colonization are not "events" that are in our past; they continue to structure our cities to the benefit of colonizing and settler (usually white) populations under the ongoing regime of racial capitalism (Moreton-Robinson 2015; Hugill 2017). Settler colonization exposes ongoing forms of political, state, and corporate violence that normalize human expendability and exploitation under climate change. Whyte (2016) writes: "Settler colonialism can be interpreted as a form of environmental injustice that wrongfully interferes with and erases the social-ecological contexts required for Indigenous populations to experience the world as a place infused with responsibilities to humans, nonhumans and ecosystems" (3). Around the world, Indigenous communities are not only more vulnerable to the impacts of a changing climate but also harmed on the front lines of energy extraction. This is the case with the Standing Rock Tribe's opposition to the Dakota Access Pipeline or the Lago Agrio oil field in Ecuador, for example. We and the other contributors in this volume recognize that most ongoing efforts at climate mitigation and adaptation are deeply entrenched in colonial and settler-colonial processes (and states) that must be directly challenged to create a just urban climate future. This is also the case for climate science itself: "Colonialism continues through the assumed universal superiority of civilized, Western ways of knowing and doing. Local and Indigenous ways of knowing are considered insufficient or simply heritage" (Liboiron 2021, 876). Some chapters in this volume will tackle this head on by articulating and challenging the ways Western interests are repeatedly centered in climate change programs and policies, while other chapters will connect the ongoing legacies of settler colonization to contemporary forms of exclusionary governance that sideline complex moral, ethical, and justice issues of climate response (see chapters by Goh, Kumar, Martizen-Lugo, and Silver, this volume).

At the same time, environmental racism constructs and naturalizes racial difference to devalue certain populations for the benefit of others (Pulido 2000, 2017). The concept of environmental justice has been widely used to describe the social and spatial processes by which low-income communities and communities of color are disproportionately likely to experience harm from environmental degradation, as well as to be excluded from environmental amenities and improvement (Walker 2012). More recently, scholars have pushed these ideas to more critically evaluate environmental (in)justice as a form of racism that requires policy intervention on the "capitalist structure[s] fueled by anti-black violence and oppression" (Ramírez 2015, 750). Wright (2021) further argues that we must better evaluate and challenge how and why Black communities are subjected to environmental violence and premature death.

Drawing on these perspectives, this book advances a critical climate justice studies lens building on theories of environmental racism that underpin the climate crisis. Stemming from foundational work by scholars of racial capitalism (Gilmore 2002; Marable 2015; Wilson 1992; Woods 2017), we understand climate injustice as inherently related to dehumanization and devaluation of human life, where inequalities among race are central to capital accumulation (Pulido 2017). The creation of devalued spaces and racialized bodies under capitalism also determine who must be "resilient" and who is safe from climate hazards. Many forms of climate-related dispossession are also produced by racial capitalism that "others," evicts, or dumps pollution on people and communities of color in the name of capital accumulation. For example, as land-grabbing accelerates globally, we are seeing it manifest as a form of racial capitalism in multiple ways. It is happening through legal frameworks for climate mitigation, as seen in the cases of biofuel production, deforestation reduction, and Reduced Emissions from Deforestation and Forest Degradation Plus (REDD+) initiatives (Parola 2020) in the Global South. Racialized dispossession also occurs as a result of urban real estate speculation and "climate-proofing" (Anguelovski et al. 2019; Keenan, Hill, and Gumber 2018), or as a form of "anticipatory ruination" to forecasted climate hazards such as rising sea levels along vulnerable coastlines in Bangladesh (Paprocki 2019). This is further complicated by racialized geopolitical expressions of power through access to land and resources, such as can be seen through both legal and forced land grabs throughout "frontier" landscapes (Enemchukwu 2019; Franco and Borras 2021). As Gonzales (2021) succinctly states: "racialized communities all over the world have borne the brunt of carbon capitalism from cradle (extraction of fossil fuels) to grave (climate change) and that a race-conscious analysis of climate change and climate displacement can reveal the commonalities among seemingly distinct forms of oppression in order to forge the alliances necessary to achieve just and emancipatory outcomes" (108).

All of this points toward a need to tackle the problem of white supremacy and white privilege head on by calling for an abolitionist climate justice (Pulido and De Lara 2018; Ranganathan and Bratman 2021). Many authors in this volume explicitly discuss racial capitalist foundations of climate injustice by analyzing historical programs and policies that have made communities of color more precarious under climate change, while others explore what antiracist ideologies would look like if they were centered in urban climate justice practice (see chapters by Knuth; Silver; Fitzgerald, Schmitz, and Stephens; Martinez-Lugo; Cox; Kumar; and Pickerill, this volume).

Finally, the persistence of heteropatriarchy excludes marginalized voices and experiences in our debates about the climate problem and its solutions, especially with regard to gender and sexuality (Israel and Sachs 2013; Tuana 2013). Of significance here is the work that feminist scholars have done related to knowledge production. Seager (2009) writes: "In the first instance, presumptions that nature can be—and should be—controlled are deeply masculinized" (14), pointing out that much of the knowledge on climate change is deeply entrenched in scientific and technical ways of knowing the problem. While science is unequivocally necessary for creating just-climate futures, it has become hegemonic as a means of defining the problem around carbon molecules, as opposed to the socioeconomic processes that create the problems, and solutions as largely technical, as opposed to political and moral (Bee, Rice, and Trauger 2015; Rice, Burke, and Heynen 2015). Others have examined how "climate change homophobia" permeates action on climate change causing harm to queer people (Gaard 2019, 94). This includes biases at climate change conferences and demonstrations. Gaard (2019) shows, for example, that "the lineup at the People's Climate March placed LGBTQ+ folks in the seventh and final group, far behind communities identified as on the 'frontlines of the crisis and the forefront of change,' . . . [which] erases the knowledges and experiences of queer and trans* people of color and reopens wounds from the larger homophobic and racist culture" (98). Authors in this volume reject these forms of heteropatriarchy by valuing the experiences and needs of female and LGBTQ+ people, as well as making visible the alternative urban climate futures already held (and being enacted) by women and queers surviving and thriving amid climate-related violence. As such, we and the contributors in this volume prioritize, value, and learn from a much broader set of knowledges and experiences than those white, patriarchal, Western ontologies that have typically dominated research and action on climate change (see chapters by Goh, Raditz, and Kumar, this volume).

Put simply, our urban spaces and lives—past, present, and future—are deeply tied to the ways that settler colonialism, environmental racism, and heteropatriarchy structure socioenvironmental relations. Elsewhere, we have defined these interlocking systems of oppression as producing a new system of "climate apartheid" (Rice, Long, and Levenda 2021). We and the contributors to this volume seek out alternative urban climate futures that reject the persistence of these ongoing acts of violence and inequalities, and especially, an intensified system of climate apartheid. We agree with Pulido and De Lara (2018) who argue for "a model of abolitionist and decolonial social movement organizing that invites interracial convergence by imagining urban political ecologies that are free of the death-dealing spaces necessary for racial capitalism to thrive" (79; see also Ranganathan and Bratman 2021 and the conclusion to this volume). Climate change has disproportionately impacted disadvantaged, marginalized, racialized, and poor people and will continue to do so unless we imagine and enact alternative urban futures. Next, we explore two conceptual vectors—security and equality—to help us think through the new possibilities for just climate futures.

Urban Futures under Climate Change: Evaluating (In)security and (In)equality

Impacts of climate change on cities have already been witnessed globally, including more intense storms, flooding, drought, and heat waves. In times of climate crisis, the idea of security has become ever more important. Climate security, generally, refers to the ways that climate change amplifies and/or creates new risks that endangers the security and safety of communities, ecosystems, economies, infrastructures, cultures, nations, and societies (Dalby 2009). These issues impact a variety of sectors and can manifest as geopolitical battles over climate-threatened resources, climate-induced migration, energy security, financial security, and more. Yet the urban context of climate security should be given more consideration as our world increasingly addresses climate change in the age of climate urbanism. Our contention is that urban climate (in)security is socially produced and unequally distributed, contributing to and exacerbating injustices.

Hodson and Marvin (2009) discuss urban ecological security as a way to frame the strategic safeguarding of flows of resources, infrastructures, and services at the urban scale, largely to maintain economic growth and (unsustainable) resource use. We use this idea to define urban climate security as a way to critically conceptualize how protection from climate hazards, resilience in response to climate catastrophes, and benefits from climate mitigation efforts are often unequally distributed in cities, creating inequities and injustices in interrelated political, social, technical, and economic systems. Several of the chapters in this book will connect the production of climate insecurity to the historical processes of oppression related to colonization, racism, and patriarchy previously described.

The idea of urban climate equality also centers on the economic, political, and social institutions that create (or inhibit) a more inclusive right to the climate-just city. We use equity in a social justice sense, where the actual causes of climate change and uneven climate vulnerability must be remade to reject (settler) colonialism, environmental racism, and heteropatriarchy. This will require modes of mitigation and adaptation that reject expendability and devaluation of human life that are produced under colonization, racism, and heteropatriarchy, as they address the ecological, political, and economic causes of climate change. More foundationally, it requires the rejection of human indispensability, full stop, across all scales of governance and all geographical contexts (Rice, Long, and Levenda 2021).

We seek to move toward more climate-secure and equitable urban futures where everyone has the ability to live happy, healthy lives, and the diversity of chapters in this volume will help materialize this goal. Charting climate-just futures requires reclaiming "security" from its militaristic and nationalist connotations, or from its use for capitalist, xenophobic, and exclusive ends. We must also reclaim "equitable" from neoliberal capitalism, which seeks to make all individuals capable of competing for scarce resources. Instead, security and equality must be reinterpreted as a social phenomenon that can direct a form of climate urbanism marked by safety, sustainability, and resilience for all—a rejection of uneven urban capitalist development and enactment of the right to the climate-just city for all.

To do this, we chart four overlapping and simultaneous, but unevenly distributed, current and future scenarios for urban climate futures that exist along the spectrums of more secure to less secure and more equitable to less equitable. Figure 0.1 provides a schematic for these current and future scenarios, as well as some examples. We name these scenarios (1) secure but inequitable, (2) insecure and inequitable, (3) insecure for all, and (4) secure and equitable. Each scenario varies in degrees of equity and justice, but they are not perfectly distinguishable, as they will exist together in dialectical tension. Yet we find this typology useful for delineating the most desirable future (scenario 4—secure and equitable) from the more alarming and harmful aspect of the other three scenarios.

We are quite aware of the drawbacks of using categories to delineate a specific set of possibilities. Yet we think that using the two concepts of (in)security and (in)equity allows academics, decision makers, and activists to better understand that we face a range of possible futures, and more importantly, we can name those we want to reject and those we want to create. We believe that the future can and should be secure and equitable (scenario 4). Authors in this volume will refer to this typology in many of their analyses to show how it can be utilized in research and activism, with the ultimate goal of moving all cities to scenario 4.

(1) SECURE BUT INEQUITABLE

The creation of more climate-secure, but still inequitable, climate futures is often a result of mainstream NGO, government, and even corporate work on climate change. These are efforts that seek to improve ecological and economic resilience to climate hazards, but do not restructure social relations that cause the problem in the first place or produce uneven vulnerability and harm. For example, one of the worst scenarios for cities in a climate-changed future is the increased existence of exclusive climate-resilient enclaves for the wealthy. Security is ensured for those in privileged positions, while vulnerability and insecurity are increased the climate-precarious. These inequities are already visible. Anguelovski et al. (2016) show how climate adaptation efforts in New Orleans and Dhaka create uneven access to flood-protective infrastructure, and how marginalized, poorer areas of Manila and Medellin are subjected to forced displacement for adaptation projects while wealthier communities are allowed to stay in place. In Kolkata the uneven geographies of climate-linked disaster risk are increasingly structured by the political economies of urbanization where new, planned suburban enclaves are resilient to flooding and other hazards, but nearby, economically interdependent slums are socially stigmatized and increasingly prone to climate-induced disasters (Rumbach 2017).

Even "green" climate solutions are continually unevenly developed, contributing to this scenario. Growing scholarship on green and low-carbon initiatives and on climate gentrification (Anguelovski et al. 2019; Rice et al. 2020; Bouzarovski, Frankowski, and Tirado Herrero 2018; Keenan, Hill, and Gumber 2018; Shokry, Connolly, and Anguelovski 2020) has shown empirical examples and given clear warning signals for a future of climate injustice manifest in the urban fabric. Eco-villages, resistant to climate collapse by virtue of their off-grid, self-reliant design, are only accessible to those with means. Major urban developments such as Eko Atlantic in Lagos combine the aesthetics of the super wealthy with ecological and resilient infrastructures. Proposed climate defense infrastructure, such as New York's Financial District and Seaport Climate Resilience Master Plan, are dedicated to protecting places with great importance to financial and political power, like Lower Manhattan. Bigger and Millington's (2020) analysis of the financial tools for resilient infrastructures show a similar pattern that contributes to increasing racialized austerity in New York and Cape Town, where "climate finance yields increasingly dangerous geographies for urbanites subject to racialized austerity and environmental change" (16). These trends forecast a future that is secure for only the wealthy and privileged, while the masses are left to struggle in increasingly vulnerable situations.

(2) INSECURE AND INEQUITABLE

Insecure and inequitable urban climate situations are currently, and will continue to be, produced as governments do nothing to address climate change—something we argue is often a purposeful political strategy to protect the status quo. This is also likely to continue to intensify nationalist and xenophobic reactions to the intense effects of climate change on marginalized people. As scholars writing on Hurricane Katrina note, the production of this disaster is rooted in a history of environmental racism and injustice that left poor communities, women, and communities of color most devastated by the storm, government response, and subsequent rebuilding of the city (Smith 2006; Bullard and Wright 2009; Jacobs 2019). The government's failure also appears in its military reaction to a socioecological disaster. The military response had roots in the Cold War–era security paradigm of distributed preparedness, popularized in cities alongside a variety of militarized defense approaches (Collier and Lakoff 2008; Graham 2009; Sassen 2017). This blurring between militarized geopolitical and ecological security is likely to be exacerbated by climate change.

Furthermore, as climate refugees and migrants meet an ever more populist, right-wing, xenophobic world, newcomers are increasingly seen as threats to both national security, racial purity, and resource abundance in an insecure and inequitable climate future. Current trends in places like drought-stricken parts of Bangladesh have caused displacement into and out of cities like Dhaka (Ayeb-Karlsson 2019), or across international boundaries, like the flow of Central American migrants toward the United States (Hallett 2019) in order to find work and secure livelihoods. While climate change is not the only driver of migration, it adds increasing complexity to a growing challenge for cities in the coming century. All of this adds to an increasingly insecure and inequitable future, especially for those who have experienced systemic injustices related to settler colonialism, racial capitalism, and heteropatriarchy.

(3) INSECURE FOR ALL

Although all aspects of security are socially constructed, there is an ephemeral leveling effect that may be presented in increasingly uncertain times. Climate hazards can impact all sorts of people, even the most well-resourced and privileged. Hurricane Sandy impacted seniors living in public housing and Wall Street bankers. In Cape Town a three-year drought led to near uniform rations of water to residents, instilling panic and concern over a "day zero" scenario. Cyclone Idai devastated nearly 90 percent of Beira, one of Mozambique's largest cities.

It is important to note that while the impacts of climate disasters can be widely felt, the moments of nature's seemingly equally distributed destruction fade almost immediately as the realities of wealth inequalities unfold. Rich people can access resources to recover more quickly. Wealthy neighborhoods are rebuilt sooner, and neighborhood changes often lead to new forms of displacement. Researchers studying Hurricane Katrina, for example, showed how the storm altered the demographic makeup of New Orleans neighborhoods, leading to changes in political representation from a mostly Black to mostly white city council and the election of the city's first white mayor in nearly four decades (Fussell 2015; Van Holm and Wyczalkowski 2019). These stark inequalities along categories of social difference—race, class, and gender—are likely to be exacerbated as the climate crisis worsens. Yet an urban climate future that is insecure for all is increasingly a reality given the dramatic nature of climate hazards.

(4) SECURE AND EQUITABLE

Despite the undesirable or even alarming aspects of the urban climate futures we have described, there are significant possibilities for alternatives. Many communities are resisting climate gentrification and displacement, and demanding more just futures. Activists are calling for systemic change to foster security and inclusivity. Across the United States, for instance, support for a Green New Deal that prioritizes the creation of security and the reduction of risk for the most vulnerable communities, often through urban adaptation projects and municipal strategies that address jobs, income inequality, housing, just energy transitions, political reform, and so on (Ocasio-Cortez 2019). Similarly, Indigenous movements have called for a Red Deal that centers the rematriation of land as a key strategy to avoid climate crisis through

Indigenous liberation and anticapitalist struggles (Estes 2019). What makes all of these efforts truly transformative is they directly name and confront the material and social processes that cause climate hazards and insecurity. For example, the Black Lives Matter movement has inspired a decolonial, antiracist agenda for just energy transitions central to more climate secure futures. As Myles Lennon (2017) writes, "we can follow BLM's energy transition agenda and cultivate relationships with the land that sustains us—the biospheric matter from which many blacks were displaced in the fossil fuel revolution—recognizing that food is both an 'energy' issue and a matter of black livelihood ... while the transition from a slave-based energy system to a fossil fuel-based energy system did not affirm the value of black lives, the transition to renewables presents new opportunities toward this end" (9–10).

In cities across the world, efforts to reclaim the city for the many are increasingly being joined with efforts toward decarbonization and resilience, in ways that highlight a right to the climate-just city and a rejection of uneven urban development. Through the struggles of these various movements, a more just and climate-secure city is emerging. As this book will showcase in the coming chapters, the efforts of activists, engaged professionals, and academics can all contribute to charting urban futures that are secure and equitable for everyone.

Contributions of This Book

The goal of this book is to provide a research and action agenda for envisioning and creating equitable and secure urban futures under climate change that are, as we previously noted, anticolonial, anti-racist, and antipatriarchal, especially from the Western perspectives and institutions in which we work and receive privilege. While many edited volumes remain purely theoretical in orientation, this book will also highlight knowledge production done with communities seeking transformative change (praxis) and meaningful learning from activist groups actually working to address the socionatural injustices of climate change impacts and policies (resistance). We recognize that theory, praxis, and resistance are always coproduced by various groups and actions. We separate the chapters into these sections, however, to highlight (1) *theory*: conceptual moves that can be made to reenvision what "climate justice" is and how insecurity and inequality are (re)produced in the climate-changed city; (2) *praxis*: efforts to meaningfully integrate climate justice into existing urban programs and policies, while expanding our understandings of what causes climate vulnerability and precarity in the first place based in real-world experience; (3) *resistance*: how grassroots groups and communities can deploy (or are already deploying) these ideas to transform on-the-ground actions related to, but also beyond, the traditional boundaries of climate justice.

The first part of the book is called "Theorizing the Just City in the Era of Climate Change." Chapters in this section examine several aspects of climate inequality under contemporary urbanization. Chapter 1 by Castán Broto, Westman, Huang, and Robin applies the principles of "just sustainabilities" to urban climate justice to argue for an "epistemic revolution" that dramatically rethinks the role of knowledge and care in our urban climate futures. In chapter 2 Shi and Bouma show that land is central to urban climate justice, especially as it relates to adaptation planning and programs. They argue that several key features of land governance must be fundamentally changed, including implementing national spatial planning, centering cooperative and community land ownership, and reorganizing intense forms of jurisdictional fragmentation. The third and fourth chapters of this section dive deeper into an emerging area of climate justice studies: green financialization and climate debt. In chapter 3 Knuth provides a thorough examination of the fiscal-financial model in U.S. cities that has produced not only uneven urban development but also racialized climate vulnerabilities in the city. She also explores lessons for urban climate justice from recent social movements to democratize local governance in U.S. cities. In chapter 4 Silver uses a case study in Mbale, Uganda, to describe and evaluate the emerging concept of green structural adjustment. This allows Silver to critically examine global center-peripheral relations of climate finance and put forth an anticolonial vision for the future of urban climate justice.

Part 2 of the book, "Climate Praxis for the Just City," begins with the perspective that creating just cities in the era of climate change will require that scholars mobilize knowledge in ways that protect vulnerable communities, challenge ongoing injustices, and reconceptualize equality in the city. We define praxis as the realization of transformative theory through practice, collaborative research, and intentional relationships between academics and others. This allows us to reimagine how our research is done and for whom. Moving beyond "community" or "participatory" planning, chapters in this section provide a vision for radical modes of doing research with and for climate-vulnerable communities in cities, based on meaningful examples from collaborative research. The first two chapters in this section evaluate and engage current climate policies and practices through more radical and inclusive lenses of climate justice. For example, chapter 5 by Fitzgerald, Schmitz, and Stephens considers the ways that climate equity measures are (or are not) incorporated into a wide range of urban plans and policies in Boston, Massachusetts, and Seattle, Washington. They also outline three primary barriers to transformative climate policies. Leichenko, Foster, and Nguyen (chapter 6) describe their experiences leading New York City's Workgroup on Community-Based Assessment of Adaptation and Equity, including the use of a coproduction model for examining multiple forms of equity in climate adaptation planning with local environmental justice organizations. The next two chapters in this section engage in reflexive research and methodologies that challenge the traditional researcher-researched divide. For example, in chapter 7 Goh draws on her own fieldwork experiences to describe "ground-up" movements for environmental justice in Jakarta, Indonesia, that draw on multilevel and multiscalar relationships among members of many differently situated groups to create new coalitions. In chapter 8 Raditz utilizes collaborative documentary filmmaking and reflexive autoethnography to render visible new forms of representational justice for climate justice and resilience of queer and trans Black and Indigenous people of color (QTBIPOC).

Part 3 focuses on "Resistance and Activism for Urban Climate Justice." Chapters here will focus on how the climate-precarious and their allies resist climate injustice and create alternative futures. This includes mechanisms of subversion for challenging status quo urban climate politics, as well as real-world efforts for integrating the needs of the most vulnerable communities into new forms of climate resilience and adaptation. The chapters in this section will discuss how and why we should learn from activist movements. Some authors do this by actually being involved in the activist movements they describe here, while others offer speculative futures for how resistance movements might reject colonial and racist social relations of climate change. For example, in chapter 9 Martinez-Lugo reflects on personal experience with grassroots organizing in the southwestern United States to imagine and enact climate justice outside of the logics of the state and capital. Martinez-Lugo examines the root causes of the climate crisis to bring understandings of the racial state and settler colonialism more centrally into focus, especially how these institutions limit the possibilities for alternative worlds. Cox (chapter 10) describes efforts by the Miami Climate Alliance—a climate and social justice coalition—to prioritize the voices of marginalized residents in the financing of climate resilience infrastructure. Cox shows how debates about municipal debt related to climate change are being recast as "key site[s] of political claims-making." In chapter 11 Kumar provides a thought-provoking story about the importance of radical friendship, hospitality, and love—especially toward those who are strangers or understood as "other"—for building an alternative and anticolonial climate politics. Lastly, Pickerill (chapter 12) critically evaluates both the limitations and possibilities that eco-communities present for more just urban climate futures, focusing specifically on the role of race and racial justice. We then end this volume with a conclusion chapter written by the editors that provides reflection on how abolition, politics of care, and reparations are necessary for building more just urban climate futures, highlighting inspiration on these topics provided by the other chapter authors. We come to these future propositions through the collective work done by all the authors included here.

REFERENCES

Anguelovski, I., Connolly, J. J., Garcia-Lamarca, M., Cole, H., and Pearsall, H. (2019). New scholarly pathways on green gentrification: What does the urban "green turn" mean and where is it going?. *Progress in Human Geography*, 43(6), 1064–1086.

Anguelovski, I., Connolly, J. J., Pearsall, H., Shokry, G., Checker, M., Maantay, J., et al. (2019). Opinion: Why green "climate gentrification" threatens poor and vulnerable populations. *Proceedings of the National Academy of Sciences*, 116(52), 26139–26143.

Anguelovski, I., Shi, L., Chu, E., Gallagher, D., Goh, K., Lamb, Z., et al. (2016). Equity impacts of urban land use planning for climate adaptation: Critical perspectives from the Global North and South. *Journal of Planning Education and Research*, 36(3), 333–348.

Attoh, K. A. (2011). What kind of right is the right to the city?. *Progress in Human Geography*, 35(5), 669–685.

Ayeb-Karlsson, S. (2019). This Bangladeshi man's story shows why linking climate change with conflict is no simple matter. *The Conversation*, December 13. https://the conversation.com/this-bangladeshi-mans-story-shows-why-linking-climate-change -with-conflict-is-no-simple-matter-127764.

Bee, B. A., Rice, J., and Trauger, A. (2015). A feminist approach to climate change governance: Everyday and intimate politics. *Geography Compass*, 9(6), 339–350.

Béné, C., Mehta, L., McGranahan, G., Cannon, T., Gupte, J., and Tanner, T. (2018). Resilience as a policy narrative: Potentials and limits in the context of urban planning. *Climate and Development*, 10(2), 116–133.

Berube, A., and Kneebone, E. (2013). *Confronting Suburban Poverty in America*. Brookings Institution Press.

Bigger, P., and Millington, N. (2020). Getting soaked? Climate crisis, adaptation finance, and racialized austerity. *Environment and Planning E: Nature and Space*, 3(3), 601–623.

Bouzarovski, S., Frankowski, J., and Tirado Herrero, S. (2018). Low-carbon gentrification: When climate change encounters residential displacement. *International Journal of Urban and Regional Research*, 42(5), 845–863.

Bouzarovski, S., and Tirado Herrero, S., 2017. The energy divide: Integrating energy transitions, regional inequalities and poverty trends in the European Union. *European Urban and Regional Studies*, 24(1), 69–86.

Bulkeley, H., and Betsill, M. 2013. *Cities and Climate Change: Urban Sustainability and Global Environmental Governance*. Routledge.

Bullard, R. D., and Wright, B. (Eds.). (2009). *Race, Place, and Environmental Justice after Hurricane Katrina: Struggles to Reclaim, Rebuild, and Revitalize New Orleans and the Gulf Coast*. Westview.

Caprotti, F. (2014). Eco-urbanism and the eco-city, or, denying the right to the city?. *Antipode*, 46(5), 1285–1303.

Castán Broto, V. (2019). The governance of climate change in urban areas. In T. Schwanen and R. van Kempen (Eds.), *Handbook of Urban Geography* (pp. 461–476). Edward Elgar.

Castán Broto, V., and Bulkeley, H. (2013). A survey of urban climate change experiments in 100 cities. *Global Environmental Change*, 23(1), 92–102.

Castán Broto, V., Robin, E., and While, A. (Eds.). (2020). *Climate Urbanism: Towards a Critical Research Agenda*. Palgrave Macmillan.

Chaturvedi, S., and Doyle, T. (2015). *Climate Terror: A Critical Geopolitics of Climate Change*. Palgrave Macmillan.

Checker, M. (2011). Wiped out by the "greenwave": Environmental gentrification and the paradoxical politics of urban sustainability. *City & Society*, 23(2), 210–229.

Cohen, D. A. (2018a). Climate Justice and the Right to the City. Current Research on Sustainable Urban Development, University of Pennsylvania. https://penniur.upenn.edu /uploads/media/Cohen.pdf.

———. (2018b). Water crisis and eco-apartheid in São Paulo: Beyond naive optimism about climate-linked disasters. *IJURR, Spotlight on Parched Cities, Parched Citizens*. https://www.ijurr.org/spotlight-on/parched-cities-parched-citizens/water-crisis-and-eco-apartheid-in-sao-paulo-beyond-naive-optimism-about-climate-linked-disasters/.

Collier, S. J., and Lakoff, A. (2008). Distributed preparedness: The spatial logic of domestic security in the United States. *Environment and Planning D: Society and Space*, 26(1), 7–28.

Cooke, T. J., and Denton, C. (2015). The suburbanization of poverty? An alternative perspective. *Urban Geography*, 36(2), 300–313.

Dahir, A. L. (2019). Africa's cities face the harshest outcomes of climate change and young people know it. *Quartz Africa*. https://qz.com/africa/1712980/climate-strike-protests-in-africa-nairobi-cape-town-lagos/.

Dalby, S. (2009). *Security and Environmental Change*. Polity.

Derickson, K. D. (2015). Urban geography I: Locating urban theory in the "urban age." *Progress in Human Geography*, 39(5), 647–657.

Dooling, S. (2009). Ecological gentrification: A research agenda exploring justice in the city. *International Journal of Urban and Regional Research*, 33(3), 621–639.

DuPuis, E. M., and Greenberg, M. (2019). The right to the resilient city: Progressive politics and the green growth machine in New York City. *Journal of Environmental Studies and Sciences*, 9(3), 352–363.

Enemchukwu, N. (2019). Resource Wars in Transformation: The Cycle of Climate Change and Its Implications in the Twenty-First Century. SSRN. https://papers.ssrn .com/sol3/papers.cfm?abstract_id=3484760.

Estes, N. (2019). A red deal. *Jacobin*, August 6. https://www.jacobinmag.com/2019/08

/red-deal-green-new-deal-ecosocialism-decolonization-indigenous-resistance
-environment.

Floater, G., Rode, P., Robert, A., Kennedy, C., Hoornweg, D., Slavcheva, R., and Godfrey, N. (2014). Cities and the new climate economy: The transformative role of global urban growth. *LSE Cities.* http://eprints.lse.ac.uk/60775/1/NCE%20Cities%20Paper01 .pdf.

Franco, J. C., and Borras Jr., S. M. (2021). The global climate of land politics. *Globalization*, 18(7), 1277–1297.

Fussell, E. (2015). The long-term recovery of New Orleans' population after Hurricane Katrina. *American Behavioral Scientist*, 59(10), 1231–1245.

Gaard, G. (2019). Out of the closet and into the climate! Queer feminist climate justice. In D. Munshi, J. Foran, K. Bhavnani, and P. A. Kurian (Eds.), *Climate Futures: Reimagining Global Climate Justice* (pp. 91–104). Zed Books.

Gilmore, R. W. (2002). Fatal couplings of power and difference: Notes on racism and geography. *Professional Geographer*, 54(1), 15–24.

Global Commission on Economy and Climate (GCEC). (2016). The Sustainable Infrastructure Imperative: Financing for Better Growth and Development. https://www .un.org/pga/71/wp-content/uploads/sites/40/2017/02/New-Climate-Economy -Report-2016-Executive-Summary.pdf.

Gonzalez, C. G. (2021). Racial capitalism, climate justice, and climate displacement. *Oñati Socio-Legal Series, Symposium on Climate Justice in the Anthropocene*, 11(1), 108–147.

Gorman-Murray, A., Morris, S., Keppel, J., McKinnon, S., and Dominey-Howes, D. (2017). Problems and possibilities on the margins: LGBT experiences in the 2011 Queensland floods. *Gender, Place & Culture*, 24(1), 37–51.

Graham, S. (2009). The urban "battlespace." *Theory, Culture & Society*, 26(7–8), 278–288.

Hallett, M. C. (2019). How climate change is driving emigration from Central America. *The Conversation*, September 6. https://theconversation.com/how-climate-change-is -driving-emigration-from-central-america-121525.

Halton, M. (2018). Climate change "impacts women more than men." *BBC News: Science and Environment*, March 8. https://www.bbc.com/news/science-environment -43294221.

Harvey, D. (2003). The right to the city. *International Journal of Urban and Regional Research*, 27(4), 939–941.

———. (2006 [1982]). *The Limits to Capital*. Blackwell.

———. (2014). *Seventeen Contradictions and the End of Capitalism*. Oxford University Press.

Hochstenbach, C., and Musterd, S. (2018). Gentrification and the suburbanization of poverty: Changing urban geographies through boom and bust periods. *Urban Geography*, 39(1), 26–53.

Hodson, M., and Marvin, S. (2009). "Urban ecological security": A new urban paradigm?. *International Journal of Urban and Regional Research*, 33(1), 193–215.

Hugill, D. (2017). What is a settler-colonial city?. *Geography Compass*, 11(5).

Israel, A. L., and Sachs, C. (2013). A climate for feminist intervention: Feminist science studies and climate change. In M. Alston and K. Whittenbury (Eds.), *Research, Action and Policy: Addressing the Gendered Impacts of Climate Change* (pp. 33–51). Springer.

Jacobs, F. (2019). Black feminism and radical planning: New directions for disaster planning research. *Planning Theory*, 18(1), 24–39.

Keenan, J. M., Hill, T., and Gumber, A. (2018). Climate gentrification: From theory to empiricism in Miami-Dade County, Florida. *Environmental Research Letters*, 13(5).

Kitchin, R. (2014). The real-time city? Big Data and smart urbanism. *GeoJournal*, 79(1), 1–14.

Kumar, M. (2018). Climate change vulnerability in urban slum communities: Investigating household adaptation and decision-making capacity in the Indian Himalaya. *Ecological Indicators*, 90, 379–391.

Lakhani, N. (2021). "'A death sentence': Indigenous climate activists denounce Cop26 deal." *Guardian*, November 16. https://www.theguardian.com/environment/2021/nov/16/indigenous-climate-activists-cop26-endangers-native-communities.

Law, T. (2019). The climate crisis is global, but these 6 places face the most severe consequences. *Time*, September 30. https://time.com/5687470/cities-countries-most-affected-by-climate-change/.

Lawton, G. (2019). The rise of real eco-fascism. *New Scientist*, 243(3243), 24.

Lennon, M. (2017). Decolonizing energy: Black Lives Matter and technoscientific expertise amid solar transitions. *Energy Research & Social Science*, 30, 18–27.

Levenda, A. M. (2019). Mobilizing smart grid experiments: Policy mobilities and urban energy governance. *Environment and Planning C: Politics and Space*, 37(4), 634–651.

Liboiron, M. (2021). Decolonizing geoscience requires more than equity and inclusion. *Nature Geoscience*, 14, 876–877.

Long, J. (2016). Constructing the narrative of the sustainability fix: Sustainability, social justice and representation in Austin, Tex. *Urban Studies*, 53(1), 149–172.

Long, J., and Rice, J. L. (2019). From sustainable urbanism to climate urbanism. *Urban Studies*, 56(5), 992–1008.

Luque-Ayala, A., and Marvin, S. (2015). Developing a critical understanding of smart urbanism. *Urban Studies*, 52(12), 2105–2116.

Marable, M. (2015). *How Capitalism Underdeveloped Black America: Problems in Race, Political Economy, and Society*. Haymarket Books.

Marshall, A. (2019). Climate protesters take to streets worldwide. *New York Times*, October 7. https://www.nytimes.com/2019/10/07/world/europe/extinction-rebellion-protests.html.

Moreton-Robinson, A. (2015). *The White Possessive: Property, Power, and Indigenous Sovereignty*. University of Minnesota Press.

Moyer, J. W., Tan, R., and Hedgpeth, D. (2019). 32 arrested as "climate rebels" shut down intersections across the district. *Washington Post*, September 20. https://www.washingtonpost.com/local/climate-change-protesters-plan-to-flood-downtown-dc-streets-monday-morning/2019/09/20/6b088e94-dbd9-11e9-89d4-37d5ffcac6f4_story.html.

National Public Radio (NPR). (2021). Uganda's Vanessa Nakate says COP26 sidelines nations most affected by climate change. https://www.npr.org/2021/11/10/1053943770/ugandas-vanessa-nakate-says-cop26-climate-summit-sidelines-global-south.

Ocasio-Cortez, A. (2019). House Resolution 109. February 7. https://www.congress.gov/bill/116th-congress/house-resolution/109/text.

O'Neill, M. S., Carter, R., Kish, J. K., Gronlund, C. J., White-Newsome, J. L., Manarolla,

X., et al. (2009). Preventing heat-related morbidity and mortality: New approaches in a changing climate. *Maturitas*, 64(2), 98–103.

Pandey, R., Alatalo, J. M., Thapliyal, K., Chauhan, S., Archie, K. M., Gupta, A. K., Jha, S. K. and Kumar, M. (2018). Climate change vulnerability in urban slum communities: Investigating household adaptation and decision-making capacity in the Indian Himalaya. *Ecological Indicators*, 90, 379–391.

Paprocki, K. (2019). All that is solid melts into the bay: Anticipatory ruination and climate change adaptation. *Antipode*, 51(1), 295–315.

Parola, G. (2020). The dangerous rise of land grabbing through climate change mitigation policies: The examples of biofuel and REDD+. *Revista de Estudos Constitucionais, Hermenêutica e Teoria do Direito (RECHTD)*, 12(3), 568–582.

Paterson, M., and Stripple, J. (2010). My space: Governing individuals' carbon emissions. *Environment and Planning D: Society and Space*, 28(2), 41–362.

Podesta, J. (2019). The Climate Crisis, Migration, and Refugees. Brookings Blum Roundtable Report. https://www.brookings.edu/research/the-climate-crisis-migration-and-refugees/.

Pulido, L. (2000). Rethinking environmental racism: White privilege and urban development in Southern California. *Annals of the Association of American Geographers*, 90(1), 12–40.

———. (2017). Geographies of race and ethnicity II: Environmental racism, racial capitalism and state-sanctioned violence. *Progress in Human Geography*, 41(4), 524–533.

Pulido, L., and De Lara, J. (2018). Reimagining "justice" in environmental justice: Radical ecologies, decolonial thought, and the Black Radical Tradition. *Environment and Planning E: Nature and Space*, 1(1–2), 76–98.

Purcell, M. (2002). Excavating Lefebvre: The right to the city and its urban politics of the inhabitant. *GeoJournal*, 58(2–3), 99–108.

Quastel, N. (2009). Political ecologies of gentrification. *Urban Geography*, 30(7), 694–725.

Ramírez, M. M. (2015). The elusive inclusive: Black food geographies and racialized food spaces. *Antipode*, 47(3), 748–769.

Ranganathan, M., and Bratman, E. (2021). From urban resilience to abolitionist climate justice in Washington, D.C. *Antipode*, 53(1), 115–137.

Rice, J. L. (2010). Climate, carbon, and territory: Greenhouse gas mitigation in Seattle, Washington. *Annals of the Association of American Geographers*, 100(4), 929–937.

———. (2014). Public targets, private choices: Urban climate governance in the Pacific Northwest. *Professional Geographer*, 66(2), 333–344.

———. (2016). "The everyday choices we make matter": Urban climate politics and the post-politics of responsibility and action. In H. Hulkeley, M. Paterson, and J. Stripple (Eds.), *Towards a Cultural Politics of Climate Change: Devices, Desires, and Dissent* (pp. 110–126). Cambridge University Press.

Rice, J. L., Burke, B. J., and Heynen, N. (2015). Knowing climate change, embodying climate praxis: Experiential knowledge in Southern Appalachia. *Annals of the Association of American Geographers*, 105(2), 253–262.

Rice, J. L., Cohen, D. A., Long, J., and Jurjevich, J. R. (2020). Contradictions of the climate-friendly city: New perspectives on eco-gentrification and housing justice. *International Journal of Urban and Regional Research*, 44(1), 145–165.

Rice, J. L., Long, J., and Levenda, A. (2021). Against climate apartheid: Confronting the persistent legacies of expendability for climate justice. *Environment and Planning E: Nature and Space.* DOI: 10.1177/2514848621999286.

Rigaud, K. K., Sherbinin, A., Jones, B., Bergmann, J., Clement, V., Ober, K., et al. (2018). Groundswell: Preparing for Internal Climate Migration, vol. 2. World Bank Group Report. http://documents.worldbank.org/curated/en/846391522306665751/Main -report.

Right to the City Alliance (RTTC). (2019). Mission and History. https://righttothecity .org/about/mission-history/.

Robin, E., and Castán Broto, V. (2020). Towards a postcolonial perspective on climate urbanism. *International Journal of Urban and Regional Research.* DOI:10.1111/1468 -2427.12981.

Romero-Lankao, P., Bulkeley, H., Pelling, M., Burch, S., Gordon, D. J., Gupta, J., et al. (2018). Urban transformative potential in a changing climate. *Nature Climate Change,* 8(9), 754.

Rumbach, A. (2017). At the roots of urban disasters: Planning and uneven geographies of risk in Kolkata, India. *Journal of Urban Affairs,* 39(6), 783–799.

Rutland, T., and Aylett, A. (2008). The work of policy: Actor networks, governmentality, and local action on climate change in Portland, Oregon. *Environment and Planning D: Society and Space,* 26(4), 627–646.

Sassen, S. (2017). When the pursuit of national security produces urban insecurity. *International Journal of Urban and Regional Research,* 34, 15–24.

Seager, J. (2009). Death by degrees: Taking a feminist hard look at the 2° climate policy. *Kvinder, Køn & Forskning.* https://doi.org/10.7146/kkf.v0i3-4.27968.

Shi, L., Chu, E., Anguelovski, I., Aylett, A., Debats, J., Goh, K., et al. (2016). Roadmap towards justice in urban climate adaptation research. *Nature Climate Change,* 6(2), 131.

Shokry, G., Connolly, J. J., and Anguelovski, I. (2020). Understanding climate gentrification and shifting landscapes of protection and vulnerability in green resilient Philadelphia. *Urban Climate,* 31, 100539.

Slocum, R. (2004). Consumer citizens and the Cities for Climate Protection campaign. *Environment and Planning A,* 36, 763–782.

Smith, N. (1984). *Uneven Development: Nature, Culture, and the Production of Space.* University of Georgia Press.

———. (2006). There's no such thing as a natural disaster. Understanding Katrina, SSRC. https://items.ssrc.org/understanding-katrina/theres-no-such-thing-as-a-natural -disaster/.

Sultana, F. (2021a). Critical climate justice. *Geographical Journal.* doi.org/10.1111/ geoj.12417.

———. (2021b). Climate and COVID-19 crises both need feminism—here's why. *The Hill,* May 16. https://thehill.com/opinion/energy-environment/553707-climate-and-covid -19-crises-both-need-feminism-heres-why.

Tozer, L. (2019). The urban material politics of decarbonization in Stockholm, London, and San Francisco. *Geoforum,* 102, 106–115.

Tuana, N. (2013). Gendering climate knowledge for justice: Catalyzing a new research agenda. In M. Alston and K. Whittenbury (Eds.), *Research, Action and Policy: Addressing the Gendered Impacts of Climate Change* (pp. 17–31). Springer.

Van Holm, E. J., and Wyczalkowski, C. K. (2019). Gentrification in the wake of a hurricane: New Orleans after Katrina. *Urban Studies*, 56(13), 2763–2778.

Walker, G. (2012). *Environmental Justice: Concepts, Evidence, and Politics*. Routledge.

While, A., Jonas, A. E., and Gibbs, D. (2004). The environment and the entrepreneurial city: Searching for the urban "sustainability fix" in Manchester and Leeds. *International Journal of Urban and Regional Research*, 28(3), 549–569.

———. (2010). From sustainable development to carbon control: Eco-state restructuring and the politics of urban and regional development. *Transactions of the Institute of British Geographers*, 35(1), 76–93.

Whitehead, M. (2013). Neoliberal urban environmentalism and the adaptive city: Towards a critical urban theory and climate change. *Urban Studies*, 50(7), 1348–1367.

Whyte, K. P. (2016). Indigenous Experience, Environmental Justice and Settler Colonialism. SSRN. http://dx.doi.org/10.2139/ssrn.2770058.

Wijsman, K., and Feagan, M. (2019). Rethinking knowledge systems for urban resilience: Feminist and decolonial contributions to just transformations. *Environmental Science & Policy*, 98, 70–76.

Wilson, B. M. (1992). Structural imperatives behind racial change in Birmingham, Alabama. *Antipode*, 24(3), 171–202.

Wolfe, P. (2006). Settler colonialism and the elimination of the native. *Journal of Genocide Research*, 8(4), 387–409.

Woods, C. (2017). *Development Arrested: The Blues and Plantation Power in the Mississippi Delta*. Verso.

Wright, W. J. (2021). As above, so below: Anti-Black violence as environmental racism. *Antipode*, 53(3), 791–809.

PART 1

Theorizing the Just City in the Era of Climate Change

CHAPTER 1

Just Sustainabilities in a Changing Climate

VANESA CASTÁN BROTO, LINDA WESTMAN,
PING HUANG, AND ENORA ROBIN

This chapter examines the limited crossover between those who look at urban climate justice in the abstract and those who engage with meaningful action on the ground (Hughes and Hoffmann 2020). We align climate justice concerns with just sustainabilities principles to advance practical initiatives to activate a just transition through social movements and socially oriented urban planning (Agyeman 2013; UN-Habitat 2016). This is an effort to develop practical responses that challenge the pervasive nature of neocolonizing practices, racism, heteronormativity, and patriarchy in urban climate action. As Aronoff et al. (2019) argue, a response to the climate crisis is only conceivable in interventions that simultaneously challenge contemporary society's structural inequalities. Climate justice must come to terms with the new dynamics of climate urbanism by engaging directly with historical questions of domination, marginalization, and oppression (Long and Rice 2019).

As highlighted in this book's introduction and by Hughes and Hoffmann (2020), just urban transitions can become a means of addressing the deepseated dynamics of environmental injustice linked to colonialism, racism, sexism, and ecological exploitation. Much of the climate change movement originates within these conditions, and hence, as it navigates an unequal society, it generates new contradictions. For example, producing renewable energy depends on the extraction of rare metals at a great environmental cost. Ensuring the safety of urban populations to climate impacts may be used as an argument to displace vulnerable populations in informal settlements, further exacerbating their precarity. Such contradictions may be salable for their emotional appeal. Such a strategy, however, will hardly help the cause of climate justice.

Acknowledging contradictions in the climate justice movement could be seen as a source of strength rather than a weakness. Contradictions point toward deep-seated structural challenges, and thus, working with contradictions may be a means of enabling progressive change (Castán Broto 2015). There is no shortage of ready-made proposals for climate action in the climate justice movement, proposals suitable for any context. The reality is, however, quite different. Any efforts to deliver sustainability in urban environments are riddled with dilemmas that inevitably impact the most disadvantaged (Metzger and Lindblad 2020). Practical experience suggests that by engaging with the complexities of the urban context, justice-inspired principles can help find collective responses for the shared challenges of climate change.

Over the years, discourses of green growth, technological optimism, and consensus-seeking policy making have presented a watered-down idea of sustainability to provide solutions for pressing problems. What those discourses do is weaken sustainability, removing its critical capacity to diagnose the fundamental contradictions we live with. Sustainability is a powerful discourse, endowed with the legacies of activists around the world who use it to demonstrate the link between social oppression and the degradation of the environment. Activists can reclaim sustainability discourses, not as a solution-oriented approach but as a critical framework with a familiar vocabulary to intervene in emancipation struggles and facilitate social change (Castán Broto and Westman 2019).

The just sustainabilities framework offers a set of four such justice-inspired principles with which we engage for this purpose (Agyeman 2013). The framework emerged from debates in the 1990s about the need to separate environmental outcomes from social justice ones. Environmental justice movements, however, demonstrated that environmental and social injustices are intertwined. Building on these debates, the just sustainabilities framework proposes four principles. The first principle is the imperative of ensuring a quality of life worth living, of dignified living conditions, for all. Not only has climate change differentiated impacts on urban populations; those impacts also compound the multiple forms of deprivation already suffered by urban dwellers in different cities and settlements. The second principle is ensuring the needs of both current and future generations, which requires tackling political representation problems. The third principle is achieving justice and equity in participation and outcomes. That is, how we design processes to facilitate action and who leads action are crucial questions from a justice perspective. The final principle entails a commitment to living within ecosystem limits. This principle is not a call for behavior change or virtue signaling. Instead, the climate justice movement needs to address current and historical inequalities and their entanglement with resource depletion and ecosystem degradation. These are not easy recipes for action; rather, just sustainabilities principles offer a revisable reference point for the collective mobilization processes that will deliver just cities in a changing climate.

Agyeman (2013) has called for further research "to assess the extent of equity and justice inputs to traditional, reform or environmental sustainability agendas in other countries, and worldwide" (185). However, until now, efforts to explain how the principles of just sustainabilities reflect (urban) climate justice concerns have been few and far between. To address this need, we argue that advancing the principles of just sustainabilities in urban environments will require a transformation akin to an epistemological revolution—a revolution that addresses both the forms of knowledge and ethics of care we bring to the question of justice.

Climate Injustice in Urban Environments

The growing visibility of climate justice discourses in mainstream media and global policy discussions imply that profound ethical challenges mire responses to climate change. For example, in the aftermath of the unlawful killing of George Floyd in Minneapolis, activist Eric Holthaus wrote that "what happened this week in Minneapolis will change the climate movement forever" because achieving racial justice is a central part of addressing the climate emergency (Holthaus 2020). He writes: "Minneapolis is one of the fastest-warming cities in the U.S. Extreme heat, a rarity even just a few decades ago, is now one of Minnesota's top three killers when it comes to extreme weather. Black people in Minneapolis, as in every U.S. city, are far less likely to have access to air conditioning, and are at a far greater risk of dying during heat waves. That's a 7-times greater increase in mortality compared to their white neighbors. I could go on and on about how the climate crisis and racism are inextricably linked."

The exploitation of Black and Brown bodies and enduring socioracial inequalities have facilitated the emergence of a capitalist, resource-intensive, global economy and now supports its maintenance. Holthaus highlighted a moment in which one death made this painfully evident. Climate change has a triple impact on people who are already disadvantaged: they suffer its impacts, they suffer the effects of climate change responses, and they suffer reduced capacity to intervene in climate change decisions (see figure 1.1). The systemic oppression regarding climate change depicted in figure 1.1 emerges from the same socioeconomic structures that created the climate crisis in the first place. Now, the response to the climate crisis may further reinforce systemic oppression. As discussed in this book's introduction, climate action may reproduce—

Lack voice in climate change–related decision-making

For example, regulatory and voluntary frameworks are poorly adapted to the needs of disadvantaged people

Bear the costs of climate change policies and benefit the least from them

For example, disadvantaged people may be more strongly impacted by different forms of carbon taxation

Suffer most from climate impacts while contributing less to carbon emissions

Structural factors shape climate change vulnerabilities and prevent people from joint structures of consumption

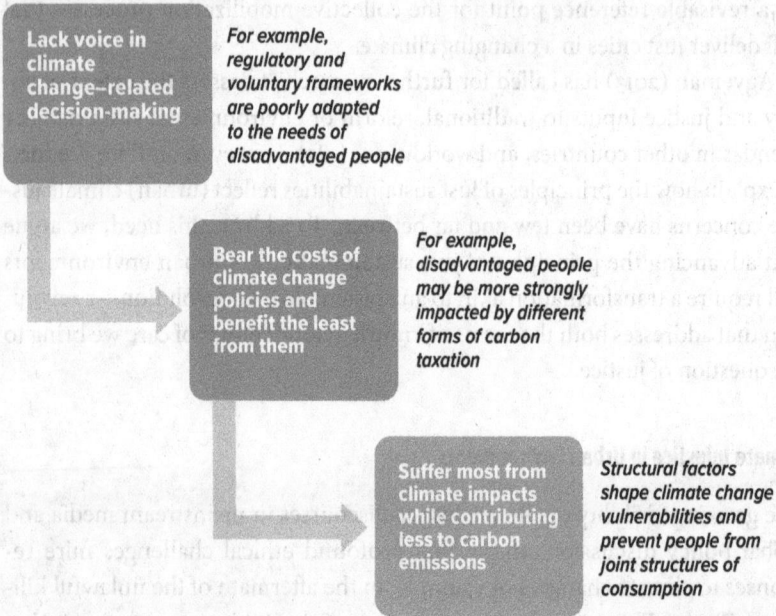

FIGURE 1.1. Low-income and disadvantaged people suffer the most from climate change and climate change action (following the analysis by Banks, Preston, and Hargreaves 2014).

rather than challenge—the systems of racial, economic, social, and gender-based oppression that have led to the current climate crisis. The task is to undo this system of oppression through truly just forms of climate action.

A key challenge for climate justice is that urban responses to climate change often exacerbate spatial, economic, and racial inequalities, for example, by creating secure, climate-proof enclaves for the economically privileged (Hodson and Marvin 2010). Analyses of climate adaptation and urban greening initiatives have long shown the inequities brought about by urban climate policies, drawing attention to the uneven distribution of benefits and harms. Interventions mobilizing large-scale (mostly private) investments for low-carbon and climate-proof infrastructures exacerbate urban inequalities (Derickson 2018). Such investments may lead to population displacements, as discussed, for instance, in Goh's (2019) study of Jakarta's defensive infrastructures against sea-level rise. Other times, infrastructures damage livelihoods, as documented, for example, in waste-to-energy conversion projects in Delhi (de Bercegol and Gowda 2019). Other low-carbon initiatives, such as building retrofitting, also lead to housing price increases and low-income population displacements, as documented in Gdansk (Bouzarovski, Frankowski, and Herrero 2018; see also

Edwards and Bulkeley 2017). As local governments address local vulnerabilities to climate impacts with limited budgets, new "greening" standards are developed in the financial sector to channel private resources into urban climate actions that reproduce existing inequalities and produce new ones (Silver 2017). Analyses of urban greening strategies in cities such as Barcelona, Medellin, and New Orleans also show that public-private partnerships to deliver green interventions open up land for private enclaves of urban green living, accessible only to privileged sectors of the population (Anguelovski, Connolly, and Brand 2018, 419). Green interventions often fail to consider "core urban issues at the intersection of racial inequalities, social and racial hierarchies, and environmental privilege" (419).

Current climate change responses may protect the most affluent urban dwellers, but the urban majority remains disproportionately vulnerable to climate change impacts—particularly low-income groups living in informal settlements. When informal settlement dwellers develop self-built alternatives for service provision, their efforts are hardly recognized (Jabeen, Johnson, and Allen 2010). Current urban responses to climate change rely on large-scale, technocratic projects instead of supporting, promoting, and building on smaller-scale grassroots efforts. Current land-use planning processes tend to exacerbate urban inequalities because they privilege elite interests (Anguelovski et al. 2016). Planning systems, for the most part, privilege economic growth and struggle to align climate action and development objectives (Chu, Anguelovski, and Carmin 2016; Rydin 2016). Thus, current climate action reinforces structural inequalities.

Implementing Just Sustainabilities in a Changing Climate

"Just sustainabilities" represent a set of principles that emphasize social and ecological well-being in tandem. It is a discourse that highlights redistributive functions to realize social justice while recognizing the need to live within the limits of our ecosystems (Agyeman, Bullard, and Evans 2003). The concept emerged in the early 2000s to respond to an "equity deficit" in sustainability and the misconception that environmental issues had to be addressed independently from social justice ones (Agyeman 2013). The concept thereby reinserted social well-being and welfare dimensions into a debate that had progressively abandoned these concerns favoring a technocratic agenda of eco-efficiency. In formulating the principles in the plural—"sustainabilities" rather than "sustainability"—the approach points to the diversity of struggles, visions, and voices involved in constructing dignified lives. Just sustainabilities is a means of reappropriating the sustainability discourse to advance radical emancipatory projects (Castán Broto and Westman 2019).

The introduction to this chapter presented four principles required to advance just sustainabilities objectives (Agyeman, Bullard, and Evans 2003; Agyeman 2013): (1) improving the quality of life and well-being; (2) meeting the needs of both present and future generations; (3) enabling justice and equity in terms of recognition, process, procedure, and outcome; and (4) living within ecosystem limits (see table 1.1). These principles are formulated in a pragmatic way that resonates with mainstream discourses because their objective is to change existing practices of urban development. However, these principles build on a history of environmental justice thinking and the environmental justice movement. They start from a redefinition of the relationship between humans and the environment that challenges preconceived ideas about the reduction of ecosystems to resources to maintain human economies. As such, they highlight the need to situate sustainability action in the context of a civil rights struggle and the pervasive nature of neocolonizing practices, racism, heteronormativity, and patriarchy. They call for the creation of international solidarities around environmental justice movements, which decenter notions of good living tied to consumerism. They call for new ways to understand the world, challenging the cognitive injustices linked to the current process of planetary destruction (de Sousa Santos 2015). In this section we offer an overview of how these principles invite reflection on climate justice.

The first principle, improving quality of life and well-being, represents a core component of the concept of sustainability. The Brundtland Report in 1987 defined sustainability as an agenda to tackle poverty, ensure access to basic services, and realize equality. The possibility of aligning economic and environmental priorities incited decades of debate, such as on the displacement of radical green agendas by prioritizing economic aims (Davison 2001; Adger et al. 2003). A just sustainabilities approach, however, refutes the notion of such trade-offs. As Agyeman (2013) explains, equating the improvement of people's lives with the degradation of the environment is a fallacy that implies that well-being only can be realized by economic expansion and exploitation of natural resources. Instead, equity in access to material resources is a starting point for delivering sustainability.

The principle of well-being and quality of life, for example, highlights the links between structures of deprivation and climate change impacts. As explained above, residents in informal settlements, who lack access to formal housing, sanitation, and energy, tend to be the populations most exposed to climate change in urban areas (Satterthwaite et al. 2018; Dodman, Archer, and

TABLE 1.1. A climate justice perspective for urban environments informed by just sustainabilities principles (own elaboration)

Just sustainabilities principles	Implications for climate justice	Considerations for research in urban environments
Improve the quality of life and well-being	▪ Recognize differentiated vulnerability, climate impacts, and distributive outcomes of climate action ▪ Consider redistributive responses to address these forms of inequality	▪ Climate change outcomes relate to cobenefits that address social development ▪ Climate injustice has historical roots ▪ Active strategies are needed to dissociate climate action from processes that deepen inequalities, such as gentrification ▪ Well-being and quality of life must be defined without prioritizing economic imperatives
Meet the needs of present and future generations	▪ Consider diverse needs, especially "marginal" perspectives (beyond a narrow interpretation from a unified white, male perspective) ▪ Include both representatives of (e.g., youth activists) and representatives for (e.g., groups acting to protect nonhuman entities)	▪ Intersectional perspectives situate vulnerabilities within specific urban histories of dispossession and oppression that extend into the past and future ▪ Multiple and conflicting views must inform any future-making experiences and strategies to address uncertainty
Achieving justice and equity in terms of recognition, process, procedure, and outcome	▪ Focus on how process-based aspects of climate justice influence outcomes ▪ From knowledge making to the delivery of solutions, the entire process needs to be open	▪ There exist multiple interpretations of who has a stake and how they can access different forums ▪ Climate justice involves paying attention to multiple forms of knowledge beyond the scientific approach tainted by its role in imperial and colonial projects
Living within ecosystem limits	▪ Develop a relationship with resources and environments that do not assume human action as a form of extraction and exploitation ▪ Consider the multiple forms of living that an environment supports	▪ Human flourishing does not depend on the tenets of capital accumulation ▪ Limits call for new symbiotic relations with our environments, something rehearsed in ideas such as "stewardship" or "care"

Satterthwaite 2019). There are many reasons for this, including that informal settlements often are located in areas prone to floods and landslides, housing is more exposed to extreme temperatures, and informal communities have higher exposure to infectious disease. Exclusion from the formal economy, including discrimination based on ethnic group, migratory status, gender, or caste, as well as from existing support systems (health care, education, social security), exacerbates vulnerability to climate risks (Satterthwaite et al. 2018).

Strategies to address these forms of deprivation must recognize the links between urban inequality and historical injustices. Patterns of affluence and poverty in cities are connected with lines of segregation produced through colonial relations and accumulation of wealth produced over generations through ownership of capital and discrimination of minority groups (Tilly 1998; Piketty 2014). For example, these links are revealed in the race wealth gap in the United States, created through a history of slavery and segregation (Baradaran 2017; Shapiro and Kenty-Drane 2005). In many urban areas around the world, racial segregation has translated into specific spatial patterns, which concentrate poverty, crime, or social marginalization in deprioritized neighborhoods (Crankshaw 2008; Harris and Curtis 2000; Logan 2014; Wilson 2007). Such residential segregation is associated with underserviced infrastructure and service provision (Acey 2007; Byrne 2012; Tighe and Ganning 2015). Economic strategies that deepen inequality, such as austerity, are similarly often constructed along racial lines (Phinney 2020; Ponder 2017), as racism follows capitalism's principles (Pulido 2016). Interventions that aim to improve the quality of life in cities must therefore work through redistribution principles. Such measures can follow welfare programs that aim for social and economic equality (such as access to education, health, and social security), or create direct access to and ownership of material resources (such as transport infrastructure, energy systems, and green space). This kind of intervention requires understanding the social distribution of the costs and impacts of climate change and how these intersect with underlying socioeconomic inequalities.

As previously noted, climate interventions in urban areas often exacerbate existing tendencies toward investment projects and urban planning to perpetuate gentrification, displacement, and exclusion. At the core of this problem lie economic rationales, quantitative indicators of progress, and an attachment to modernity and technocratic ideals. A just sustainabilities approach requires decentering urban development agendas toward nonmaterial conditions of well-being (including social and psychological dimensions), non-Western definitions of needs and human rights (by drawing on philosophies from elsewhere, such as Buen Vivir), prosperity for nonhuman beings (such as through recognition of nature to live and thrive), and collective aspirations for dignified life conditions (for example, community-based metrics of well-being).

The second principle, meeting the needs of both present and future generations, has also been part of sustainable development since its introduction. The Brundtland Report states that "humanity has the ability to make development sustainable to ensure that it meets the needs of the present without compromising the ability of future generations to meet their own needs" (Brundt-land et al. 1987, chap. 2). In its weakest form, this oft-repeated definition takes the form of eco-efficiency agendas enacted on behalf of people's consumption in the future. However, meeting the needs of present and future generations requires recognizing the inherent uncertainty about the future: what will happen there, and what will the future generations' needs be? Engaging with tomorrow's needs entails considering a diversity of knowledges and experiences: "If we assume that the preferences of future peoples are unknown, we must decide instead how to realize political representation for a generation of people that do not yet exist. This dilemma calls for the need to reimagine planning as a process that brings together a multiplicity of voices, confronts and reflects on these views together, and provides people the right to decide today and in the future" (Castán Broto and Westman 2019, 131).

Thus, the question of representation is crucial to achieving climate justice. Representation means considering the different experiences of climate policy and climate change impacts. Representation also entails considering the voices of activists willing to look beyond consensus to examine current action's emerging contradictions. Groups and individuals can stand for a perspective, such as in the representation of youth by young individuals. The youth strikes of Fridays for Future represent future generations within the climate justice movement (Taylor, Watts, and Bartlett 2019). Groups and individuals can also act for a perspective, for example, through activist groups advocating for nature's interests (O'Neill 2001). In any case, the commitment to representation involves a messy process with multiple actors in an increasingly complex debate occasionally renewed, for example, by activists (Edelman 2020). A key element of recognition, for example, is to provide antidiscrimination strategies to tackle subtle forms of structural and institutional discrimination, such as those learned from anti-racist activists (Pascoët and Siklossy 2020). Concrete interventions may include affirmative action in environmental institutions, implicit bias training, screening laws and policies to understand mechanisms of exclusion, or critically examining memorializing structures influencing accounts of environmental history, such as museums (Pascoët and Siklossy 2020).

The third principle of just sustainabilities relates to enabling justice and equity in terms of recognition, process, procedure, and outcome. This principle is directed toward transcending the false dichotomy between outcomes and process. Environmental justice outcomes cannot be achieved at the expense of procedural compromises, for example, excluding certain groups of the population from decision-making. At the same time, a fair process that never delivers outcomes would eventually lose the interest of everyone involved. This principle combines insights from the different dimensions of justice introduced, among others, by Nancy Fraser (1995). Output/distributive notions of justice (which relate to equality in access to resources and mechanisms of distribution) need to be considered alongside procedural dimensions (which refer to participation and ability to engage in and shape political processes of decision-making), and dimensions of recognition (which builds on theories of structural discrimination based on identities such as gender, ethnicity, race, ability, class, or sexual orientation) (Castán Broto and Westman 2019).

Relating the third just sustainabilities principle to urban climate justice highlights several issues. In terms of outcomes, it brings us back to the inequitable burden of climate change on socially marginalized groups as discussed above. Environmental racism entails siting conflicts but also broader questions about who participates in environmental decision-making (Bullard et al. 2008). The legacies of environmental racism have become painfully visible in the wake of the Black Lives Matter movement. Resource extractivism always has a direct impact on somebody's life, somewhere. Climate change has further distributed those impacts, reaching vulnerable populations at a distance via carbon emissions. Environmental burdens (justice in outcome) are directly related to deficits in participation (justice in procedure), as well as to recognition (discrimination rooted in political identity).

The international policy framework for climate change provides limited opportunities for citizens' participation in the diplomatic setting of the United Nations Framework Convention for Climate Change, initially devised as a mechanism for interstate negotiation and target setting. This setting exacerbates the exclusion of groups already underrepresented in political processes, such as women (Olson 2014) and Indigenous peoples (Schroeder 2010). Other scales of action may offer more opportunities for participation. However, even in contexts where local decision makers make explicit efforts to create participatory processes, barriers to inclusion arise from the framing of discussions to practical aspects such as the assumptions about the availability of time and economic resources (Gibson-Wood and Wakefield 2013).

Long-standing concerns about discrimination of groups and communities based on sociocultural and political identity (Young 1990; Fraser 1995) are central to urban climate justice. There is also an increasing understanding that recognition entails epistemic justice—the revaluation of multiple forms of knowledge (de Sousa Santos 2015). Actions for climate justice will have to be situated in a local context, build on lived experience, and bring direct benefits to citizens' everyday lives. Such approach requires challenging common ways of practice and considering climate adaptation and mitigation measures that build on Indigenous and local knowledges, integrating different cosmologies and understandings of human-ecological relations (Codjoe, Owusu, and Burkett 2014; Ebhuoma and Simatele 2019; Kosoe, Diawuo, and Osumanu 2019).

The final principle refers to living within ecosystem limits. The concept of ecosystems limits has long been linked to fears about the depletion of environmental resources (e.g., Meadows et al. 1972). The concept of planetary boundaries has revived this debate. Planetary boundaries define a "safe operating space for humanity" by calculating thresholds that, if transgressed, would launch abrupt and irreversible large-scale environmental change (Steffen et al. 2015). A focus on quantitative evaluation and extrapolation of current consumption trends into the future is common in these approaches. There is a risk that we move from planetary boundaries to prescriptions for human life. Just sustainabilities recognize that we, as humans, live in a finite world and are entirely dependent on and interconnected with nature, but just sustainabilities proves distinct from problematic Malthusian approaches by assigning equal importance to recognizing that the concept of "limits" is a social construction that depends on our perceptions of scarcity and abundance.

Linking the notion of ecosystem limits back to urban climate justice leads to two main points. The first is the need to consider "limits" alongside social justice and social needs (Agyeman 2013). This involves viewing limits from the perspective of deep inequalities inherent in the historical accumulation of environmental deterioration. For example, industrialized countries bear historical responsibility for a large share of greenhouse emissions, but the impacts of climate change are disproportionally felt in nations with very low accumulated emissions (Dellink et al. 2009; Althor, Watson, and Fuller 2016). In the broader context of exploitation of environmental resources, ecofeminists and environmental activists have argued that countries of the North owe ecological debt toward countries of the South, whose resources and population were exploited through colonialism (Alier 1997, 2003; Isla 2007). Specific tools have been developed to account for "carbon debt," including the greenhouse development rights framework, which calculates carbon budgets based on historically accumulated emissions and the need for further development (Baer et al. 2009). Despite decades of negotiations, there is still no effective mechanism to ensure that the industrialized nations compensate for such debt.

The second point relates to the need to think of meaningful forms of abundance to reduce emissions. Capitalism creates perceptions of scarcity through the production of enclosures and reinforcing the need for consumption (Kallis and March 2015). However, we can rethink abundance in noneconomic terms by linking it to variety, personal relations, experience, and meaning. There are numerous proposals for the revaluation of cultures of sharing, cooperation, communal systems, and comanagement (Kallis and March 2015) and emotional attachment to material objects through cultures of care and maintenance (Eisenstein 2011). In the context of urban climate policy, these proposals tackle practices and values of high carbon consumption without reducing climate action to a consumption act that we can address by buying the right kind of product, such as an electric vehicle or a more efficient heating system.

Conclusion

The year 2020 was unique and challenging: a global pandemic, lockdowns that created additional tensions and vulnerabilities, and the international growth of the Black Lives Matter movement after the death of George Floyd in Minneapolis have raised high expectations. Is this a moment of reckoning with our human existence on earth? Again, the impacts of the pandemic have been suffered by those who were already exposed to risks as well as by those who were at the front line of the crisis, as caregivers, doctors, cleaners, drivers, deliverers, factory and logistics workers—those whose lives were already burdened by multiple layers of oppression. Again, we have seen populist responses from unmentionable leaders in governments who struggled to see any merit in the possibility of providing collective responses to a public health crisis and who have spoken unashamedly of the need to maintain the dynamics of capital accumulation above the preservation of human life. These have been myopic analyses and myopic responses that fail to engage in any meaningful way with deeper concerns for the future. However, these responses have not always found an echo within populations of concerned citizens who understand that individual and collective welfare are intrinsically linked. The COVID-19 pandemic has made more urgent than ever the need to deliver alternative visions of societies that can respond together and protect its members whatever their situation.

   With climate change, these dilemmas will become more visible. In urban environments, the impacts of climate change are forcing decisions that require moving our moral muscles and are well embedded within discourses of environmental justice. That triad of aspects in figure 1.1 (exposure and vulnerability to climate change impacts, exposure and vulnerability to climate policy, and participation in the current policy-making process) points toward the interlinked nature of justice questions. Hughes and Hoffmann (2020) have argued that a just urban transition needs to engage with three dimensions: urban planning and politics, justice frameworks and principles, and transition strategies. Such an approach raises research questions about accountability, agency, capacity, and diversity. Our engagement with just sustainabilities discourses has shown that two aspects are particularly complicated and will require something akin to an epistemological revolution:

1. The reliance on Western systems of knowledge and performance measuring systems that reduce the scope of action to that which can be measured in a positivist sense or that which can be associated with a tangible benefit.
2. The lack of empathy to understand ecological systems' autonomy, as our environment continues to be portrayed as a resource to be exploited for our benefit.

While the just sustainabilities principles link climate justice to actions on the ground that we know already are effective in advancing different aspects of justice, some of those prescriptions continue to be perceived as unachievable by those who still embrace technocratic fantasies of unlimited growth. However, recent social movements have started to connect the dots between racial and social injustices, the intrinsic unfairness of capitalist economies, and the climate crisis. Historical injustices cannot be wished away. Advancing climate justice in urban environments is within reach for concerned policy makers, activists, and local communities, but there are thick epistemological walls that so far prevent it.

REFERENCES

Acey, C. (2007). Space vs. race: A historical exploration of spatial injustice and unequal access to water in Lagos, Nigeria. *Critical Planning*, 14, 49–70.

Adger, W. N., Brown, K., Fairbrass, J., Jordan, A., Paavola, J. Rosendo, S., and Seyfang, G. (2003). Governance for sustainability: Towards a "thick" analysis of environmental decisionmaking. *Environment and Planning A*, 35(6), 1095–1110.

Agyeman, J. (2013). *Introducing Just Sustainabilities: Policy, Planning, and Practice.* Zed Books.

Agyeman, J., Bullard, R. D., and Evans, B. (Eds.). (2003). *Just Sustainabilities: Development in an Unequal World.* MIT Press.

Agyeman, J., Schlosberg, D., Craven, L., and Matthews, C. (2016). Trends and directions in environmental justice: From inequity to everyday life, community, and just sustainabilities. *Annual Review of Environment and Resources*, 41, 321–340.

Alier, J. M. (1997). Deuda ecológica y deuda externa. *Ecología Política*, 14, 157–173.

———. (2003). *Deuda ecológica: ¿quién debe a quién?.* Icaria Editorial.

Althor, G., Watson, J. E. M., and Fuller, R. A. (2016). Global mismatch between greenhouse gas emissions and the burden of climate change. *Scientific Reports*, 6, 20281.

Anguelovski, I., Connolly, J., and Brand, A. L. (2018). From landscapes of utopia to the margins of the green urban life: For whom is the new green city?. *City*, 22, 417–436.

Anguelovski, I., Shi, L., Chu, E., Gallagher, D., Goh, K., Lamb, Z., et al. (2016). Equity impacts of urban land use planning for climate adaptation: Critical perspectives from the Global North and South. *Journal of Planning Education and Research*, 36, 333–348.

Aronoff, K., Battistoni, A., Aldana Cohen, D., and Riofrancos, T. (2019). *A Planet to Win: Why We Need a Green New Deal*. Verso.

Baer, P., Kartha, S., Athanasiou, T., and Kemp-Benedict, E. (2009). The greenhouse development rights framework: Drawing attention to inequality within nations in the global climate policy debate. *Development and Change*, 40(6), 1121–1138.

Banks, N., Preston, I., and Hargreaves, K. (2014). *Climate Change and Social Justice: An Evidence Review*. Joseph Rowntree Institution.

Baradaran, M. (2017). *The Color of Money: Black Banks and the Racial Wealth Gap*. Harvard University Press.

Bouzarovski, S., Frankowski, J., and Tirado Herrero, S. (2018). Low-carbon gentrification: When climate change encounters residential displacement. *International Journal of Urban and Regional Research*, 42, 845–863.

Brundtland, G. H., Khalid, M., Agnelli, S., Al-Athel, S., and Chidzero, B. (1987). *Our Common Future*. World Commission on Environment and Development.

Bullard, R. D. (1999). Dismantling environmental racism in the USA. *Local Environment*, 4(1), 5–19.

Bullard, R. D., Mohai, P., Saha, R., and Wright, B. (2008). Toxic wastes and race at twenty: Why race still matters after all of these years. *Environmental Law*, 38, 371–411.

Byrne, J. (2012). When green is white: The cultural politics of race, nature, and social exclusion in a Los Angeles urban national park. *Geoforum*, 43(3), 595–611.

Castán Broto, V. (2015). Contradiction, intervention, and urban low carbon transitions. *Environment and Planning D: Society and Space*, 33(3), 460–476.

Castán Broto, V. and Westman, L. (2019). *Urban Sustainability and Justice: Just Sustainabilities and Environmental Planning*. ZED Books.

Chu, E., Anguelovski, I., and Carmin, J. (2016). Inclusive approaches to urban climate adaptation planning and implementation in the Global South. *Climate Policy*, 16, 372–392.

Codjoe, S. N. A., Owusu, G., and Burkett, V. (2014). Perception, experience, and indigenous knowledge of climate change and variability: The case of Accra, a sub-Saharan African city. *Regional Environmental Change*, 14(1), 369–383.

Crankshaw, O. (2008). Race, space and the post-Fordist spatial order of Johannesburg. *Urban Studies*, 45(8), 1692–1711.

Davison, A. (2001). *Technology and the Contested Meanings of Sustainability*. SUNY Press.

de Bercegol, R., and Gowda, S. (2019). A new waste and energy nexus? Rethinking the modernisation of waste services in Delhi. *Urban Studies*, 56, 2297–2314.

Dellink, R., Den Elzen, M., Aiking, H., Bergsma, E., Berkhout, F., Dekker, T., and Gupta, J. (2009). Sharing the burden of financing adaptation to climate change. *Global Environmental Change*, 19(4), 411–421.

Derickson, K. D. (2018). Urban geography III: Anthropocene urbanism. *Progress in Human Geography*, 42, 425–435.

de Sousa Santos, B. (2015). *Epistemologies of the South: Justice against Epistemicide.* Routledge.

Dodman, D., Archer, D., and Satterthwaite, D. (2019). *Responding to Climate Change in Contexts of Urban Poverty and Informality.* SAGE.

Ebhuoma, E. E., and Simatele, D. M. (2019). "We know our Terrain": Indigenous knowledge preferred to scientific systems of weather forecasting in the Delta State of Nigeria. *Climate and Development,* 11(2), 112–123.

Edelman, S. (2020). Who stands up for ecology? The politics of sustainable land use in Stockholm. In J. Metzger and J. Lindblad (Eds.), *Dilemmas of Sustainable Development* (pp. 39–49). Routledge.

Edwards, G. A. S., and Bulkeley, H. (2017). Urban political ecologies of housing and climate change: The "coolest block" contest in Philadelphia. *Urban Studies,* 54, 1126–1141.

Eisenstein, C. (2011). *Sacred Economics: Money, Gift, and Society in the Age of Transition.* North Atlantic Books.

Fraser, N. (1995). From redistribution to recognition? Dilemmas of justice in a "postsocialist" age. In S. Seidman and J. C. Alexander (Eds.), *The New Social Theory Reader* (pp. 188–196). Routledge.

Gardner, F., and Greer, S. (1996). Crossing the river: How local struggles build a broader movement. *Antipode,* 28(2), 175–192.

Gibson-Wood, H., and Wakefield, S. (2013). "Participation," white privilege, and environmental justice: Understanding environmentalism among hispanics in Toronto. *Antipode,* 45(3), 641–662.

Goh, K. (2019). Urban waterscapes: The hydro-politics of flooding in a sinking city. *International Journal of Urban and Regional Research,* 43, 250–272.

Harris, F. R., and Curtis, L. A. (2000). *Locked in the Poorhouse: Cities, Race, and Poverty in the United States.* Rowman & Littlefield.

Heiman, M. K. (1996). Race, waste, and class: New perspectives on environmental justice. *Antipode,* 28, 111–121.

Hodson, M., and Marvin, S. (2010). Urbanism in the Anthropocene: Ecological urbanism or premium ecological enclaves?. *City,* 14(3), 298–313.

Holthaus, E. (2020). I don't know how to explain to you that people live here. *The Correspondent,* June 4.

Hughes, S., and Hoffmann, M. (2020). Just urban transitions: Toward a research agenda. *Wiley Interdisciplinary Reviews: Climate Change,* 11(3), 640.

Isla, A. (2007). An ecofeminist perspective on biopiracy in Latin America. *Signs: Journal of Women in Culture and Society,* 32(2), 323–332.

Jabeen, H., Johnson, C., and Allen, A. (2010). Built-in resilience: Learning from grassroots coping strategies for climate variability. *Environment and Urbanization,* 22(2), 415–431.

Kallis, G., and March, H. (2015). Imaginaries of hope: The utopianism of degrowth. *Annals of the Association of American Geographers,* 105(2), 360–368.

Kosoe, E. A., Diawuo, F., and Osumanu, I. K. (2019). Looking into the past: Rethinking traditional ways of solid waste management in the Jaman South Municipality, Ghana. *Ghana Journal of Geography,* 11(1), 228–244.

Logan, J. R. (2014). *Separate and Unequal in Suburbia*. Census brief prepared for US2010, Brown University, https://s4.ad.brown.edu/Projects/Diversity/data/report/report 12012014.pdf.

Long, J., and Rice, J. L. (2019). From sustainable urbanism to climate urbanism. *Urban Studies*, 56, 992–1008.

Meadows, D. H., Meadows, D. L., Randers, J., and Behrens, W. W. (1972). The Limits to Growth. Potomac Associates, New York.

Metzger, J., and Lindblad, J. (Eds.). (2020). *Dilemmas of Sustainable Urban Development: A View from Practice*. Routledge.

Olson, J. (2014). Whose voices matter? Gender inequality in the United Nations Framework Convention on Climate Change. *Agenda*, 28(3), 184–187.

O'Neill, J. (2001). Representing people, representing nature, representing the world. *Environment and Planning C: Government and Policy*, 19(4), 483–500.

Pascoët, J., and Siklossy, G. (2020). *Intersectional Discrimination in Europe: Relevance, Challenges and Ways Forward*. Report by the Center for Intersectional Justice (CIJ) commissioned by the European Network Against Racism.

Phinney, S. (2020). Rethinking geographies of race and austerity urbanism. *Geography Compass*, 14(3), e12480.

Piketty, T. (2014). *Capital in the Twenty-First Century*. Harvard University Press.

Ponder, C. S. (2017). *The life and debt of great American cities: Urban reproduction in the time of financialization*. University of British Columbia, PhD thesis.

Pulido, L. (2016). *Flint, Environmental Racism, and Racial Capitalism*. Taylor & Francis.

Rydin, Y. (2016). Sustainability and the financialisation of commercial property: Making prime and non-prime markets. *Environment and Planning D: Society and Space*, 34, 745–762.

Satterthwaite, D., Archer, D., Colenbrander, S., Dodman, D., Hardoy, J., and Patel, S. (2018). Responding to climate change in cities and in their informal settlements and economies. International Institute for Environment and Development, paper prepared for the IPCC for the International Scientific Conference on Cities and Climate Change in Edmonton, March.

Schlosberg, D., and Collins, L. B. (2014). From environmental to climate justice: Climate change and the discourse of environmental justice. *Wiley Interdisciplinary Reviews: Climate Change*, 5, 359–374.

Schroeder, H. (2010). Agency in international climate negotiations: The case of indigenous peoples and avoided deforestation. *International Environmental Agreements: Politics, Law and Economics*, 10(4), 317–332.

Shapiro, T. M., and Kenty-Drane, J. L. (2005). The racial wealth gap. In *African Americans in the U.S. Economy* (175, 177). Rowman & Littlefield.

Silver, J. (2017). The climate crisis, carbon capital and urbanisation: An urban political ecology of low-carbon restructuring in Mbale. *Environment and Planning A*, 49, 1477–1499.

Steffen, W., Richardson, K., Rockström, J., Cornell, S. E., Fetzer, I., Bennett, E. M., et al. (2015). Planetary boundaries: Guiding human development on a changing planet. *Science*, 347(6223), 1259855.

Taylor, M., Watts, J., and Bartlett, J. (2019). Climate crisis: 6 million people join latest wave of global protests. *Guardian*, September 27.

Tighe, J. R., and Ganning, J. P. (2015). The divergent city: Unequal and uneven development in St. Louis. *Urban Geography*, 36(5), 654–673.

Tilly, C. (1998). *Durable Inequality*. University of California Press.

UN-Habitat (2016). World Cities Report 2016: Urbanisation and Development, Emerging Futures. UN-Habitat.

Wilson, D. (2007). *Cities and Race: America's New Black Ghetto*. Routledge.

Young, M. I. (1990). *Justice and the Politics of Difference*. Princeton University Press.

CHAPTER 2

Reclaiming Land Governance under Climate Change

LINDA SHI AND DIETRICH BOUMA

After a decade of mainstreaming urban climate adaptation, scholars and practitioners are finally waking up to the reality that many adaptation projects and processes reproduce historic dynamics of inequality (Rice, Levenda, and Long, this volume). At the individual property level, adaptation projects tend to be implemented by wealthier residents who can afford to elevate their homes or buy "climate-proofed" new construction; at the neighborhood level, adaptation projects like floodable parks or seawalls can directly displace existing residents or gentrify beneficiary communities; at the metropolitan level, wealthier cities are better poised to write and receive grants and implement adaptive projects, often at the expense or opportunity cost of less powerful jurisdictions. These adaptation and resilience-planning efforts, while forward looking from an environmental perspective, typify efforts to preserve processes and institutions that have contributed to social inequality and environmental dispossession and exploitation (Henrique and Tschakert 2020).

Conflict over land (e.g., who has rights to land, where, for what purpose) and land governance institutions (e.g., individual- or community-level property rights, taxation, and financialization; governmental rules of incorporation, land use planning, and zoning; and territorial strategies of economic and spatial development) lies at the core of these debates. Accelerating climate impacts, both direct and indirect, are destabilizing core institutions structuring territorial development, land administration, land-based finance and taxation, and property rights, all of which are predicated on stable environmental conditions. Yet while the issue of land is a central challenge for adaptation and the focus of research in geography, agriculture, and law, most adaptation discourse addresses the processes of contestation and technologies of management rather than land governance. Advocates of adaptation justice have similarly focused on contesting outcomes—for example, whether seawalls should prioritize frontline communities, whether communities should rebuild after a disaster—but not the institutions shaping municipal land scarcity or why safe, affordable housing is so constrained.

This chapter argues that we need to track climate vulnerability to its roots in the urbanization of capital and the institutions governing land as part of redressing systemic inequity and injustice in climate urbanism. As summarized in table 2.1, this chapter reviews four aspects of land governance that contribute to present-day conditions of urban climate vulnerability and how they, in turn, are impacted by climate change. First, the vulnerability of urban centers to climate change derives in part from hyperurbanization, sociospatial inequality, and attendant processes of territorial socioecological transformation. Second, urban administration typically fragments land into jurisdictions and hierarchies that enable the extraction of resources from rural hinterlands to urban centers. The establishment of boundaries is central to defining geographies of wealth distribution, particularly under conditions of climate-exacerbated resource scarcity. Third, urban development is commonly predicated on land-based finance, whether through property taxes, land leases, or development capitalization, such that climate impacts create crises of capital accumulation that trigger searches for a new spatial fix. Finally, property rights regimes in cities in most countries privilege individual ownership, reflecting classist and racist processes of land titling. Here too, changing ecological characteristics of land induce new conflicts over rights, responsibilities, and scales of ownership.

In short, climate change injects new sources of uncertainty and instability into the institutions underlying capitalist processes of accumulation and social reproduction. This raises questions of land—who should have rights to "climate-safe" land, whether cities can continue to rely on land-based finance, and who should make decisions about land resources and development. Domains related to land governance therefore are becoming chief sites of contestation that may result in existing institutions doubling down to shore up wealth accumulation or creating more progressive possibilities of radical change. In a post-2020 era, local governments, adaptation professionals, and equity advocates are voicing their commitment to combating historic racism and classism. In support of these efforts, the chapter concludes with a call for researchers, practitioners, and activists interested in advancing a different adaptation politics to grapple with and reimagine alternative institutions of land governance.

TABLE 2.1. Institutions shaping land governance and their relationship to climate vulnerability

	Institutional mechanism	Impact on cities' climate vulnerability	How climate destabilizes institutions
Urbanization of capital	Processes of hyper-urbanization: • consolidate landownership • delink resource sources and sinks from where most people live, consume, produce • increase inequitable access to land	• Unsustainable natural resource management worsens exposure to "natural" hazards • Concentrated human settlement in climate-vulnerable places and inadequate water for that geography • Dense urban environments are inflexible to change	• Climate change redraws map of lands that are comparatively more or less desirable for human development • Centers of capitalism among most vulnerable • Climate migration is changing historic trends in urbanization and territorial development
Territorial fragmentation	• Establishment of boundaries divides land into independent jurisdictions, each with the capacity to exclude people or externalize costs	• Depending on level of fragmentation, can impede coordination at ecological or territorial scales • Uneven capacity among jurisdictions and presence of borders enables formation of ecological enclaves and climate slums	• Climate impacts cut across boundaries • Uncoordinated climate adaptation reduces regional competitiveness and well-being • New emphasis on importance of regional scales of governance
Land fiscalization	• Land-based finance drives land conversion for development maximization regardless of environmental conditions	• Pressures to develop at-risk areas worsens exposure and adaptive capacity • Uneven fiscal capacity historically created communities with uneven levels of infrastructure preparedness	• Direct and indirect climate impacts reduce municipal land-based revenues leading to widespread defaults and larger-scale fiscal crises
Land privatization	• Land privatization helps maximize development and externalize impacts • Dispossession and titling processes can deepen unequal landownership	• Erosion of collective/communal institutions requires individual scale of adaptation • Unequal land ownership translates into unequal assets for mobility and migration under climate change	• Climate impacts shift ownership in perpetuity into usufruct rights • Neighborhood or collective scales needed to support adaptation • Conflicts over land use regulations and liability may redefine property rights bundles

Urbanization of Capital

The built environment covers only 1 percent of the world's land mass but supports 55 percent of the world's population (United Nations 2016). Over 20 percent of the world's population lives in one of 512 cities with more than one million residents. The most industrialized countries have urbanization levels around 75–85 percent, and urbanization in the Global South is on track to reach these levels in future decades. These demographic realities reflect what

Harvey (1978) has described as the "urbanization of capital." The raw materials necessary for production and city building require the transformation of nature into land and commodities, the enclosure of rural land to create a cheap land-dispossessed labor force, and concentration of land among fewer landowners to enable industrialized production and extraction (Cronon 1992; Polanyi 1944). Cities grow as industrialization and capitalism require the co-location of people for production and consumption. The built environment becomes both the space that enables the social reproduction of this system and an investment outlet for accumulated capital.

The mirror image of hyperurbanization is the erosion of population, human capital, social institutions, and traditional socioecological relationships in rural geographies. The resultant and growing urban income and housing inequality is well known: the top 10 percent of global wealth holders have 82 percent of the world's wealth, while the bottom 50 percent owns less than 1 percent (Credit Suisse 2019), but land inequality is central to the story too. In South Asia and Latin America, the top 10 percent of landowners control 75 percent of agricultural land value while the bottom 50 percent owns less than 2 percent (Bauluz, Govind, and Novokmet 2020). In the United States, the top five agricultural landowners (all white) own more than all African Americans combined, and white Americans account for 96 percent of owners, 97 percent of land value, and 98 percent of acreage (Gilbert, Wood, and Sharp 2002).

All of these processes commenced and continue to varying degrees under colonial and postcolonial regimes of racial capitalism. In settler-colonial countries like the United States and Australia, colonial powers dispossessed native communities of land and its resources, waged genocide, exploited labor, and extracted raw materials to fuel capitalist expansion (Veracini 2015). In urban spaces across Africa, colonial spatial segregation created racial, ethnic, and sociocultural enclaves with uneven urban development for colonizer and colonized (Njoh 2007). The perpetuation of these economic and social forces continues to divest rural communities of land and resources, livelihoods, and well-being, and to draw ever more people into uneven, segregated cities.

What does this mean for societal vulnerability and justice under climate change? First, the concentration of people in major urban centers makes urban residents especially exposed to environmental change. Historic inequality in land, housing, and water provision and unsustainable urban resource management make low-income, structurally disadvantaged groups and new migrants especially vulnerable. Urban density and runaway land prices challenge efforts to develop affordable housing or engage in managed retreat. Second, rural residents are also impacted by changing climates, but many lack control over land or other assets to combat these changes. Large corporate landowners in the Sundarbans are more likely to convert rice paddies to shrimp farming, salinating the water and soil much more than rising seas (Paprocki 2020). National and local governments often seize rural water resources or land for floodwater storage to ensure their own resilience (Hommes et al. 2019; Shi et al. 2021). Land and livelihood dispossession, compounded by slow and sudden onset environmental hazards from climate change, further propel internal rural-urban and small city to megacity migration (Rigaud et al. 2018). The effect is a greater concentration of people on a small area of land. Third, climate impacts and migration add to the strain on the social welfare programs, entitlement programs, infrastructures, and affordable housing in cities. With cities already struggling to provide services, especially under austerity capitalism, it becomes an open question how, and if, they can sustain this under climate change. A system that first strips people of means of self-sufficiency, then concentrates them without providing adequate wages, land, housing, or services ensures systemic vulnerability to climate impacts.

This spatial development model bears within itself deep contradictions and seeds of crisis under climate change. For the global financial elite, climate change threatens entire global supply chains, economic command and control centers, and their corresponding global real estate markets. Their concern can be seen in the global Task Force on Climate-Related Financial Disclosures tasking its eight-hundred-plus members holding $118 trillion in assets to disclose their climate risks and in central and public banks beginning to categorize assets as "green" or "brown" based on their climate impacts or risks (Keenan 2019). This process of disclosures provides the basis of revaluing assets and realigning market values to internalize climate costs, which in turn sends market signals to redirect flows of capital spatially to new geographies. Through a spatial fix under climate change, neoliberal capitalism could once again prove remarkably resilient to global shocks and stresses (Harvey 2001; Taylor 2020). However, local governments, residents, and the built environment are not as mobile or agile. This contradiction between use value and exchange value of the built environment creates significant potential for new landscapes of boom and bust. Administrative boundaries play a critical role in creating "secure but inequitable" enclaves (Rice, Levenda, and Long, this volume) that define rights to resources and sites of comparative resilience.

Administrative Fragmentation

The delineation of administrative boundaries to manage the urbanization of capital reflects historic contestation between state and private sector factions, competing local and political identities, as well as racial, ethnic, and other social divisions. The United States is famously fragmented—each metropolitan region has on average one hundred municipal governments with over twenty-five thousand local governments nationwide. While early central cities grew by annexation, growing suburbs rejected annexation, preferring instead to incorporate as independent administrations that could establish their own zoning, taxation, and redistributive policies. The growth in incorporated U.S. local governments coincided with the rollout of segregationist policies, such that newly established local governments could adopt zoning regulations that directly excluded Black residents (and less overtly, Jews, Hispanics, Catholics, Asians, and others) or economically excluded them through minimum lot or housing size requirements (Glotzer 2020). Regardless of debates as to whether administrative fragmentation promotes government efficiency and competitiveness, these boundaries have helped exclude disadvantaged groups, concentrating where they reside to specific geographies (such as central cities or more recently inner-ring suburbs), and defining who benefits from local wealth redistribution.

In other parts of the world, urban administrations tend to be more centralized, although cities like Metro Manila in the Philippines or Dakar, Senegal, following recent decentralization reforms are similarly fragmented. In other cases, extensive urbanization has created fragmented megaregions that are only sometimes restructured into capital cities or governorates. In China the central government famously established domestic mobility controls through hukou registration, providing residents social services and rights to public benefits only in the location of their parents' birthplace, a status that is inherited (Bach 2010; Chan 2019). Land tenure status (more on this below) similarly creates boundaries of formal areas worthy of investment and infrastructure upgrades, and informal areas that are legally deemed unworthy of public resources. Thus, boundaries of administration, broadly construed, contribute to the enclavization of wealth and, conversely, the confinement of poverty to select geographies, with different governments and politics determining the level of redistribution.

One ramification of this for cities' climate vulnerability is that the instruments of land use planning chronically concentrated the urban poor and racial minority groups in the least environmentally desirable land within municipalities (Mohai, Pellow, and Roberts 2009; Stephens 1996). For instance, in New York City, 19 percent of the city's public housing run by the city's housing authority is in the hundred-year floodplain (Furman Center 2013) and 70 percent are within former redlined areas (Winkler 2016). This was due to the city siting public housing on public land, much of it coastal or redlined, without advocating for dispersed or integrated housing. Recent studies also show that redlined neighborhoods have the fewest street trees and the least open space and are the most exposed to urban heat islands and rising temperatures, particularly compared to single-family homes on large, leafy lots in surrounding suburban cities (Hoffman, Shandas, and Pendleton 2020). Economic zoning that required larger lot sizes and setbacks excluded African Americans in U.S. cities, and their policy transfer to the Global South has had an even more pronounced effect. Chronic underproduction of housing and the exclusion of affordable housing from safer geographies relegate the newest immigrants to the least desirable land, often along waterways, landfills, shorelines, and wetlands. As the environmental justice literature repeatedly attests, environmentally hazardous plants, industries, landfills, and more are often sited in these locations as well (Taylor 2014). Environmentally hazardous sites are likely to follow the poor as they change locations due to climate gentrification. Climate injustices compound persistent environmental injustices.

Administrative boundaries serve to isolate risks and economic fallout within these jurisdictions. Historically, following cycles of economic boom and bust, central cities housed the urban poor when suburbs became desirable, then inner-ring suburbs became impoverished as central cities revitalized. Under climate change, municipalities face uneven levels of exposures to climate impacts. The creation of socially exclusionary enclaves in turn translates into differentiated capacity to collectively organize on behalf of investments in climate-resilient infrastructure, disaster response and recovery services, insurance, or remedies for the loss of private property (Anguelovski et al. 2016; Mach et al. 2019). Feedback loops for shrinking economic competitiveness and land value of cities on the front lines of climate change reduce government revenues and capacity to respond. As climate change induces more mobile households to migrate (Hauer 2017), low-income groups currently occupying more climate-desirable land are more likely to be displaced to climate-vulnerable municipalities (Aune, Gesch, and Smith 2020).

Land Fiscalization

Land-based development has become a principal source of revenue for many local governments worldwide. Experts of fiscal policy say property taxes are the most ideal source of municipal finance. It least distorts market incentives, closely aligns local actions with the rewards of good governance, and are best collected at the local level as compared to sales, income, and business taxes that span multiple jurisdictions (Bird 1998). Property taxes are one of the largest sources of revenue for many (former) Commonwealth countries of the Global North. The combination of administrative fragmentation that barred regional tax sharing and local tax reliance for municipal revenues consigned minority-majority, postindustrial, and many smaller communities to lower-quality services, schools, housing, infrastructure, and environmental conditions. Non–property tax revenues can also be reliant on growth and development. For instance, China collected land lease fees by expropriating rural farmland and selling to developers to fund local government infrastructure and housing development (Rithmire 2017). In most of the Global South, property taxes are an insignificant source of revenues, but budget reliance on port activity, foreign direct investment, and intergovernmental transfers similarly rely on urban land development. Raising revenues from these sources or local user fees and charges requires either a larger base against which to levy charges or wealthier residents who can pay higher fees. As a result, municipal budgets and local officials are highly dependent on development maximization, especially as decentralization and fiscal austerity at higher levels of government push service responsibilities to the lowest levels of government (Ahmad et al. 2005).

What are the implications of this for urban experiences of climate vulnerability? Historically, there has been little relationship between a geography's exposure to "natural" hazards and its local development planning (Burby 1998). Even today, municipalities have little incentive to reduce development in high hazard areas and little recourse if significant proportions of land become inundated or are otherwise undevelopable due to extreme heat and drought. Studies consistently find that even with more acute awareness of climate impacts on disaster frequency and intensity, local plans are uncoordinated and place new development and infrastructure in areas their own plans find are chronically at risk of flooding (Berke et al. 2018; Malecha, Woodruff, and Berke 2020). Cities' fiscal reliance on growth therefore continues to create new geographies of risk. For instance, even as Lagos, Manila, and Jakarta evict informal settlements from along canals, they also build new smart cities and high-end subdivisions on reclaimed land on the waterfront (Ajibade 2019; Goh 2019). The fiscalization of land, coupled with administrative fragmentation, creates uneven landscapes of wealth and fiscal capacity within and across metropolitan regions to adapt to climate impacts. The concentration of poverty, affordable housing, and marginalized social groups can translate into lower tax revenues and government adaptive capacity in places where these groups reside.

Reliance of capital and government systems on land commodification and fiscalization places these institutions on a collision course with global environmental changes that alter the long-term productivity and profitability of current landholdings and administrative jurisdictions. The more urbanized and concentrated that urbanization is—such as in primate cities like Mexico City, Bangkok, Jakarta, and Cairo—the greater the national exposure to urban impacts and fallout from climate change. To some extent, extensive financialization of land and risk have mitigated this impact of these transformations. Taylor (2020) finds that the growth of insurance and reinsurance markets has absorbed some of the shocks. However, like the oceans' ability to absorb carbon, these too have limits to what they can absorb.

Land Privatization

Private property rights have been central to capitalism and wealth accumulation. Traditional property rights around the world encompassed a variety of models, with the proportion of land held publicly, privately, or communally varying according to economic, cultural, social, geographic, ecological, legal, and political factors. Research on the commons suggests collectives are capable of sustainably managing common property resource pools under specific conditions (Larson and Bromley 1990; Ostrom 1990). However, these models do not maximize profit in the short term or funnel profit into individual or state coffers (von der Osten, Kirley, and Miller 2017). By contrast, European thinkers have advanced private property regimes that allow individuals to maximize property value and exchange property in the marketplace (De Soto 2000; Demsetz 1967). These concepts have spread globally through processes of colonization and developmentalism, including through multilateral development agencies and trade liberalization treaties (McMichael 2012). Colonial and state actions have undermined communal land holdings and their social and land governance institutions that resisted both state and market forces (Curry and McGuire 2002; Tsing 2005). Yet processes of privatization have not necessarily translated into greater social justice or environmental health. As Larson and Bromley wrote, the structure of private property rights does not eliminate interdependence among landowners but enables users to ignore the cost of externalities to others (Larson and Bromley 1990). The role of the state to coordinate these contradictions to achieve socially and environmentally optimal outcomes is no more assured than under common pool resource regimes.

Informal settlements, in this context, have become a conceptual figure ground of all the uses and users left out of the formal system. Informality results from countries' unwillingness or inability to redistribute wealth and provide adequate housing and infrastructure for the poor (Roy 2005). Their informality paradoxically justifies continued eviction or marginalization. Informal settlement dwellers not only lack services but also live under the continuous threat of being displaced for new formal housing and businesses and the road, water, and transportation systems that support them (Gupte et al. n.d.; Roberts and Okanya 2020). Under climate change, informal settlers are more exposed to environmental impacts, water scarcity or service suspension, and disasters (Satterthwaite et al. 2020), even as their informality becomes a new justification for climate adaptation efforts to remove urban poor residents in order to securitize the formal sector (Ajibade 2019; Goh 2019; Paprocki 2020).

Yet climate change is also unleashing new rounds of contestation among property owners, financial corporations, and public agencies that may redefine the rights and responsibilities associated with private property. For one, who should bear the burden of individual property loss? Relocating vulnerable people is costly and opposed by targeted communities. Wealthier groups receive infrastructure investments rather than eviction notices (Anguelovski, Irazábal-Zurita, and Connolly 2019). Insurance could minimize risk and provide compensation in case of disaster, but flood insurers left flood risk to governments long ago, and reinsurers may not be able to handle the level of impact under climate change (Taylor 2020). Still other coalitions are mobilizing lobbying efforts to convince national governments to redistribute tax dollars into major infrastructure projects that chiefly project downtown business districts (Ajibade 2017; Teicher 2018). These are all examples of efforts to "dump risk" and shift responsibility for who should pay for losses or to maintain the status quo. In the United States, the fact that the government does not reimburse homeowners for many other causes of property devaluation—such as deindustrialization, capital or white flight, or the siting of undesirable infrastructure—suggests that continued maintenance of public infrastructure underwriting coastal development or public compensation for disaster-induced losses may change in the future. The magnitude of the infrastructure maintenance backlog and fiscal crisis in many countries, especially after 2020, will likely accelerate these debates, particularly in smaller and less wealthy municipalities.

For another, the need for neighborhood-scale responses to climate adaptation reopens the value of community, place-based organizations. The scale of individual property rights poses a major impediment to widespread action, as it requires each household to navigate climate impacts, adaptation options, insurance plans, and relocation options, and coordinating their timing, financing, and implementation. Urbanization and private property rights regimes have typically dissolved community institutions that once held decision-making power, managed cooperative enterprises, pooled resources, and provided mutual assistance (Shi et al. 2018). In informal settlements, community development initiatives, such as Community Organizations Development Institute, Self-Employed Women's Association, and Slum Dwellers International, have sought to reconstruct the "commons" to help informal settlers survive. Such structures provide a template for re-commoning.

The Problem of Land for Climate Adaptation and Pathways Forward

Tracking climate vulnerability back to its roots shows how the current crisis derives from centuries-long processes of changing relationships between human societies and land. This process has entailed the enclosure of the commons through war, colonialism, genocide, or trade agreements; boundary drawing to create political jurisdictions with differential authority and capacity to exclude or eject less desirable residents; and the privatization and fiscalization of land to maximize economic productivity. As table 2.1 summarizes, these institutional legacies shape societies' climate vulnerability by concentrating human settlements, accentuating income and land inequality within and across cities, and eroding ecosystem health to withstand climate impacts. The reliance on land-based finance itself renders municipalities vulnerable to climate change, while private property rights and fragmented governance have eroded collective capacity to engage in system adaptation.

At the same time, climate change—among many other crises—destabilizes institutions established under periods of relative climatological stability. For now, mainstream adaptation reflects efforts by capital to ensure its resilience in global cities (e.g., through lobbying for protective infrastructure), risk dumping, and changing standards in financial industries to minimize exposure and seek new sites for future investment. These strategies seek to sustain existing institutions for land governance and reproduce processes of wealth accumulation. A growing community of scholars and activists recognize that resilient development is replicating the same dynamics that contributed to past rounds of dispossession, racism, and exclusionary development. Urban renewal now takes place with green infrastructure instead of highways, climate gentrification replaces segregation, and the logic of "bluelining" (where banks stop lending to flood-prone areas) replaces redlining by race. Yet the severity of the crisis for capital and governments as well as emergent social movements for climate justice also point to new processes of contestation in both physical and institutional realms.

As an example, Southeast Florida epitomizes how land governance has contributed to climate vulnerability and constrains equitable and sustainable institutional responses. For twelve thousand years, the Seminole inhabited what is now the southeastern United States, living within the region's ecosystem limits. After the last of three wars against the Seminole ended in 1856, the U.S. government transferred half of Florida's land to the state government. Land rich and money poor, Florida sold land to developers and train magnates in return for railroad and road construction (Carter 1974). Following major floods in the 1920s and 1940s, Florida and the U.S. Army Corps of Engineers built major drainage projects, including a perimeter levee along the eastern edge of the Everglades. This dramatically extended land for agriculture and urban development in Southeast Florida, whose growth helped make Florida one of the fastest-growing U.S. states post–World War II.

With growth came increasingly severe environmental impacts, from droughts and wildfires in the Everglades, to pollution, congestion, and groundwater salinization. In response, municipalities in Southeast Florida led the state to enact sweeping reforms throughout the 1960s to the 1980s. New policies required the state and municipalities to develop comprehensive land-use plans, gave states and regions the ability to review lower-level plans, set aside large sums to purchase environmentally endangered lands and recreation areas, and created a suite of new environmental and planning regulations (Carter 1974). In later decades, the state passed additional regulations requiring hazard mitigation, avoiding development in high hazard areas, and requiring sufficient infrastructure (Deyle, Chapin, and Baker 2008).

These policies, however, did not challenge land use fiscalization, rules of municipal incorporation, or individual property rights in vulnerable areas. As the region grew, new suburban developments increasingly incorporated into their own cities to control tax rates and budgets, set zoning requirements that protected local neighborhood character (by race and class), and exclude unwanted others. As of 2017, Southeast Florida tallied 108 municipalities. In Miami-Dade County, 28 out of 34 municipalities have populations under fifty thousand; the smallest has 86. The state does not levy income tax (to better attract retirees), and local governments rely on property taxes and user fees and charges to fund increasingly state-devolved service obligations. This incentivizes cities to maximize development within their boundaries, even in high hazard areas with development restrictions (Deyle, Chapin, and Baker 2008).

Southeast Florida's much-lauded Climate Change Compact avidly champions climate adaptation but also supports continued growth (150 percent in the next fifty years) in a region that lies entirely within twenty-five feet of sea level. Cities compete for resilience grants and climate-aware investors, each seeking to burnish their resiliency credentials, build out their last bit of high ground, and keep capital flowing until reinsurance markets pop the real estate bubble. Cities are upzoning inland and upland neighborhoods, which will likely displace Black communities who lack collective authority to manage growth pressures or resist gentrification. Meanwhile, homeowners have sued local governments both for adaptation investments that curtail private rights (such as beach renourishment that made waterfronts public property) and inaction (such as not repairing coastal roads that are repeatedly washed out).

At the root of Florida's challenges lies an extractive mode of land governance that has never looked beyond fifty years, much less one hundred. The fragmentation of land into severable cities and parcels, each seeking to maximize property values within their respective boundaries, is a colonial system that embeds logics of exclusion and extraction into the DNA of land governance. Such a system has contributed to present levels of inequality and unsustainability and challenges any notion of collective action, regional coordination, and rights of nature.

How, then, do we build a just city that equitably offers all residents security under climate change (Rice, Levenda, and Long, this volume)? If land governance tools have driven inequality and climate vulnerability, then climate and social justice cannot be achieved without reforming land governance. Just land governance under climate change requires constraining capital-driven market forces and creating greater space for public and communal action at multiple scales of land policy—from national planning to property parcel.

Reviving national spatial planning to steer development away from primate cities or global cities can help reduce housing and land shortages and, by connection, extreme levels of capital accumulation. This is not a call for sprawl but a recognition that spatial planning for integrated regional, urban, and rural development can support climate and housing justice. Regionalized development can also reduce exposure to concentrated climate risks as well as urban demands on overtapped environmental systems. By investing in social development and infrastructure, national governments could also encourage migration pathways toward urban areas in safer geographies with less overheated housing markets. For example, proposals in Bangladesh suggest creating college towns in small cities and subsidizing education for young women to encourage voluntary migration away from the Sundarbans. Youth from coastal agricultural families could gain a professional education and eventually bring aging families from coastal communities to join them, rather than migrating to Dhaka to join ranks of the poorest, newest migrants (Ghosh, Danda, and Bandyopadhyay 2016; Huq 2019). Academics have proposed a "Green (Rust) Belt and Road Initiative" that shifts U.S. development from California, Arizona, and the South toward the Rust Belt through national investments in a green economy (Dolšak and Prakash 2019). These somewhat blue-sky visions nevertheless provide examples of how to articulate the spatial implications of radical visions for just transitions (Climate Justice Alliance n.d.). Spatial reorganization, however, by no means ensures greater social equity or justice, as new development in neglected regions can still rehash past cycles of dispossession and exclusion (Paprocki 2020).

To counteract historic top-down or market-driven oppression, restorative community landownership policies at the national or subnational levels must permit and prioritize cooperative land and enterprise ownership that enables residents to accumulate wealth rather than being displaced by it. Centuries of government policies have created a framework for accumulation through land, and alternative policies to make possible thriving communities must also systematically legalize, protect, prioritize, legitimate, and fund cooperative models. Many European nations, New Zealand, and socialist democratic nations have achieved this within a modern, industrialized, and capitalist system. Research shows that cooperatives and community land trusts are effective in wealth redistribution and resisting displacement where government policies enable them to do so (Moore and McKee 2012). The United States, long a bastion of individual business and housing ownership, is witnessing a renaissance of collectively owned enterprises, community land trusts, and even cooperatively owned mobile home parks (Spicer 2020; French, Giraud, and Ward 2008). In the Philippines, the Community Mortgage Program permits informal settlements to acquire land cooperatively with support of national housing finance. Like many community-led settlement upgrading networks, it is predicated on land tenure security and basic services that are the basis for adaptive capacity in the face of climate change (Satterthwaite et al. 2020). Land readjustment—a strategy primarily used in Asia to reorganize development for post-disaster recovery, slum upgrading, or transit-oriented development—could also support managed retreat (Hong and Tierny 2018). These and other strategies focused on land rights for renters, informal settlers, and low-income residents are necessary to counteract historic marginalization (Boone and Agyeman 2020; Van Sant 2019).

Finally, challenges for municipal governance and fiscal viability under climate change present an opportunity to steer municipalities toward cooperation and away from fragmentation. Most metropolises across the Global North and South have piloted soft-power regional cooperation, while others have revived interest in watershed management for flood risk reduction, building on support for ecosystems and nature-based adaptation (Shi 2020, 2019). While beneficial, harder regionalism may be necessary. For example, the San Francisco Bay Area has passed a regional tax to raise funding for wetland restoration that would, among other things, mitigate flood risks. The Regional Plan Association of New York, New Jersey, and Connecticut has advocated something similar, creating a regional Coastal Adaptation Commission that would charge taxes and raise revenues for an Adaptation Trust Fund (RPA 2017). Given the scale of local fiscal exposure to climate impacts from wildfires and floods, longer-term climate impacts will raise questions about administrative boundaries. Since climate impacts do not respect administrative boundaries, regionalism can provide more equitable adaptation distribution and provide efficiency gains.

Conclusion

These alternative institutional and managerial tools, strategies, and policies have always been available, so the real question is what kind of politics would mobilize behind progressive reforms. Rising populist backlash to the discontents of globalization has produced a crisis of legitimacy in the contemporary neoliberal world order, and potentially a policy window wherein alternatives (of left and right varieties) enter the mainstream (Chu and Shi, 2021). The political conflicts between white supremacy and Black Lives Matter in the United States underscore the challenges ahead. Nevertheless, leftist social movements are pushing back on racialized and discriminatory discourse to articulate new visions for social and climate justice. For advocates and practitioners, this chapter argues that mobilization efforts would be more effective if they attended to the institutional drivers shaping current government, private sector, and individual responses and constraining possible alternatives. For academics, this chapter suggests that these spaces are important spheres in which to identify and evaluate emerging climate conflicts, as well as to synthesize knowledge in support of policy and activist mobilization.

REFERENCES

Ahmad, J., Devarajan, S., Khemani, S., and Shah, S. (2005). Decentralization and Service Delivery. World Bank Policy Research Working Paper 3603. World Bank.

Ajibade, I. (2017). Can a future city enhance urban resilience and sustainability? A political ecology analysis of Eko Atlantic City, Nigeria. *International Journal of Disaster Risk Reduction*, 26, 85–92.

———. (2019). Planned retreat in Global South megacities: Disentangling policy, practice, and environmental justice. *Climatic Change*, 157(2), 299–317.

Anguelovski, I., Irazábal-Zurita, C., and Connolly, J. T. (2019). Grabbed urban landscapes: Socio-spatial tensions in green infrastructure planning in Medellín. *International Journal of Urban and Regional Research*, 43(1), 133–56. https://doi.org/10.1111/1468-2427.12725.

Anguelovski, I., Shi, L., Chu, E., Gallagher, D., Goh, K., Lamb, Z., et al. (2016). Equity impacts of urban land use planning for climate adaptation: critical perspectives from the Global North and South. *Journal of Planning Education and Research*, 36(3), 333–348. https://doi.org/10.1177/0739456X16645166.

Aune, K., Gesch, D., and Smith, G. (2020). A spatial analysis of climate gentrification in Orleans Parish, Louisiana post-hurricane Katrina. *Environmental Research*, 185 (June), 109384. https://doi.org/10.1016/j.envres.2020.109384.

Bach, J. (2010). "They come in peasants and leave citizens": Urban villages and the making of Shenzhen, China. *Cultural Anthropology*, 25(3), 421–458. https://doi.org/10.1111/j.1548-1360.2010.01066.x.

Bauluz, L., Govind, Y., and Novokmet, F. (2020). Global Land Inequality. World Bank. https://wid.world/document/global-land-inequality-world-inequality-lab-wp-2020-10/.

Berke, P. R., Malecha, M. L., Yu, S., Lee, J., and Masterson, J. H. (2018). Plan integration for resilience scorecard: Evaluating networks of plans in six U.S. coastal cities. *Journal of Environmental Planning and Management*, June 1–20. https://doi.org/10.1080/09640568.2018.1453354.

Bird, R. M. (1998). Local and regional revenues: Realities and prospects. In R. M. Bird and F. Vaillancourt (Eds.), *Perspectives on Fiscal Federalism*, (pp. 1–31). World Bank, Washington D.C.

Boone, K., and Agyeman, J. (2020). Land loss has plagued Black America since Emancipation—is it time to look again at "Black commons" and collective ownership? *The Conversation*, June 18. https://theconversation.com/land-loss-has-plagued-black-america-since-emancipation-is-it-time-to-look-again-at-black-commons-and-collective-ownership-140514.

Burby, R. J. (1998). *Cooperating with Nature: Confronting Natural Hazards with Land Use Planning for Sustainable Communities*. National Academy Press.

Carter, L. J. (1974). *The Florida Experiment: Land and Water Policy in a Growth State*. Johns Hopkins University Press.

Chan, K. W. (2019). China's Hukou System at 60: Continuity and reform. In Y. Yep and J. Wang (Eds.), *Handbook on Urban Development in China* (pp. 59–79). Edward Elgar.

Chu, E., and Shi, L. (2022). Urban Climate Adaptation: Discontents and Alternative Politics. In J. Sowers, S. D. VanDeveer, & E. Weinthal (Eds.), *The Oxford Handbook of Comparative Environmental Politics*. Oxford University Press. https://doi.org/10.1093/oxfordhb/9780197515037.013.27.

Climate Justice Alliance. (n.d.). Climate Justice Alliance. https://climatejusticealliance.org/.

Credit Suisse. (2019). Global Wealth Report 2019. Credit Suisse Research Institute, Switzerland.

Cronon, W. (1992). *Nature's Metropolis: Chicago and the Great West*. W. W. Norton.

Curry, J. M., and McGuire, S. (2002). *Community on Land: Community, Ecology, and the Public Interest*. Rowman & Littlefield.

Demsetz, H. (1967). Toward a theory of property rights. *American Economic Review*, 57(2), 347–359.

De Soto, H. (2000). *The Mystery of Capital*. Basic Books.

Deyle, R. E., Chapin, T. S., and Baker, E. J. (2008). The proof of the planning is in the platting: An evaluation of Florida's hurricane exposure mitigation planning mandate. *Journal of the American Planning Association*, 74(3), 349–370.

Dolšak, N., and Prakash, A. (2019). Jobs and climate change: America's (rust) belt and road initiative. *Forbes*, July 14. https://www.forbes.com/sites/prakashdolsak/2019/07/14/jobs-and-climate-change-americas-rust-belt-and-road-initiative.

French, C., Giraud, K., and Ward, S. (2008). Building wealth through ownership: Resident-owned manufactured housing communities in New Hampshire." *Journal of Extension*, 46(2). https://archives.joe.org/joe/2008april/a3.php.

Furman Center. (2013). Sandy's Effects on Housing in New York City. Fact Brief. http://furmancenter.org/research/publication/fact-brief.

Ghosh, N., Danda, A. A., and Bandyopadhyay, J. (2016). *Away from the Devil and the Deep Blue Sea: Planned Retreat and Ecosystem Regeneration as Adaptation to Climate Change*. WWF-India.

Gilbert, J., Wood, S., and Sharp, G. (2002). Who owns the land? Agricultural land ownership by race/ethnicity. *Rural America*, 17(4), 55–62.

Glotzer, P. (2020). *How the Suburbs Were Segregated: Developers and the Business of Exclusionary Housing, 1890–1960*. Columbia University Press. http://cup.columbia.edu/book/how-the-suburbs-were-segregated/9780231179997.

Goh, K. 2019. Urban waterscapes: The hydro-politics of flooding in a sinking city. *International Journal of Urban and Regional Research*, 43(2), 250–272. https://doi.org/10.1111/1468-2427.12756.

Gupte, J., Rao, V. K., McGregor, A., and Lakshman, R. (2019). Global Report on Internal Displacement: Demolition, Forced Evictions and Wellbeing in the City. Institute of Development Studies, University of Sussex. https://www.internal-displacement.org/global-report/grid2019/downloads/background_papers/Jaideep_FinalPaper.pdf.

Harvey, D. (1978). The urban process under capitalism: A framework for analysis. *International Journal of Urban and Regional Research*, 2(1–3), 101–131. https://doi.org/10.1111/j.1468-2427.1978.tb00738.x.

———. (2001). Globalization and the spatial fix. *Geographische Revue*, 2, 23–30.

Hauer, M. (2017). Migration induced by sea-level rise could reshape the U.S. population landscape. *Nature Climate Change*, 7(5), 321–325. https://doi.org/10.1038/nclimate3271.

Henrique, K. P., and Tschakert, P. (2020). Pathways to urban transformation: From dispossession to climate justice. *Progress in Human Geography*, October. https://doi.org/10.1177/0309132520962856.

Hoffman, J. S., Shandas, V., and Pendleton, N. (2020). The effects of historical housing policies on resident exposure to intra-urban heat: A study of 108 U.S. urban areas. *Climate*, 8(1), 12. https://doi.org/10.3390/cli8010012.

Hommes, L., Boelens, R., Harris, L. M., and Veldwisch, G. J. (2019). Rural-urban water struggles: Urbanizing hydrosocial territories and evolving connections, discourses and identities. *Water International*, 44(2), 81–94. https://doi.org/10.1080/02508060.2019.1583311.

Hong, Y.-H., and Tierny, J. (2018). Global Experiences in Land Readjustment. UN-HABITAT, Nairobi. https://unhabitat.org/sites/default/files/documents/2019–05/global_experiences_in_land_readjustment_urban_legal_case_studies_volume_7_1.pdf.

Huq, S. (2019). Developing Climate Resilient Migrant Friendly Towns in Bangladesh to Tackle Future Climate Migration. Presented at the Cornell University Department of City and Regional Planning Colloquium Lecture Series, Ithaca, New York, September 27.

Keenan, Jesse M. (2019). A climate intelligence arms race in financial markets. *Science*, 365(6459), 1240–1243. https://doi.org/10.1126/science.aay8442.

Larson, B. A., and Bromley, D. W. (1990). Property rights, externalities, and resource degradation: Locating the tragedy. *Journal of Development Economics*, 33(2), 235–262.

Mach, K., Kraan, C., Hino, M., Siders, A. R., Johnston, E., and Field, C. (2019). Managed retreat through voluntary buyouts of flood-prone properties. *Science Advances*, 5(10), 8995. https://doi.org/10.1126/sciadv.aax8995.

Malecha, M. L., Woodruff, S. C., and Berke, P. R. (2020). Planning to mitigate, or to exacerbate, flooding hazards? Evaluating a Houston, Texas, network of plans in place during Hurricane Harvey using a plan integration for resilience scorecard." *Natural Hazards Review*, 22(4), 04021030.

McMichael, P. (2012). Instituting the development project. In P. McMichael (Ed.), *Development and Social Change: A Global Perspective*, 5th ed. (pp. 26–54). SAGE.

Mohai, P., Pellow, D., and Roberts, J. T. (2009). Environmental justice. *Annual Review of Environment and Resources*, 34(1), 405–430.

Moore, T., and McKee, K. (2012). Empowering local communities? An international review of community land trusts. *Housing Studies*, 27(2), 280–290. https://doi.org/10.1080/02673037.2012.647306.

Njoh, A. J. (2007). *Planning Power: Town Planning and Social Control in Colonial Africa*. Routledge. https://www.routledge.com/Planning-Power-Town-Planning-and-Social-Control-in-Colonial-Africa/Njoh/p/book/9781138978539.

Ostrom, E. (1990). *Governing the Commons: The Evolution of Institutions for Collective Action*. Cambridge University Press.

Paprocki, K. (2020). The climate change of your desires: Climate migration and imaginaries of urban and rural climate futures. *Environment and Planning D: Society and Space*, 38(2), 248–266.

Polanyi, K. (1944). *The Great Transformation: The Political and Economic Origins of Our Time*. Beacon Press.

Rigaud, K. K., de Sherbinin, A., Jones, B., Bergmann, J., Clement, V., Ober, K., et al. (2018). Groundswell: Preparing for Internal Climate Migration. World Bank.

Rithmire, M. E. (2017). Land institutions and Chinese political economy: Institutional complementarities and macroeconomic management. *Politics and Society*, 45(1), 123–153.

Roberts, R. E., and Okanya, O. (2020). Measuring the socio-economic impact of forced evictions and illegal demolition: A comparative study between displaced and existing informal settlements. *Social Science Journal*, 59(1), 119–138.

Roy, A. (2005). Urban informality: Toward an epistemology of planning. *Journal of the American Planning Association*, 71(2), 147–158.

Regional Plan Association (RPA). (2017). Coastal Adaptation: A Framework for Governance and Funding to Address Climate Change. A Report of the Fourth Regional Plan. Regional Plan Association, New York.

Satterthwaite, D., Archer, D., Colenbrander, S., Dodman, D., Hardoy, J., Mitlin, D., and Patel, S. (2020). Building resilience to climate change in informal settlements. *One Earth*, 2(2), 143–156.

Shi, L. (2019). Promise and paradox of metropolitan regional climate adaptation. *Environmental Science and Policy*, 92 (February), 262–274.

———. (2020). Beyond flood risk reduction: How can green infrastructure advance both social justice and regional impact? *Socio-Ecological Practice Research*, 2(4), 311–320.

Shi, L., Ahmad, F., Shukla, P., and Yupho, S. (2021). Shared vulnerability, splintered solidarity: Governing water across urban-rural divides. *Global Environmental Change*, 70, 102354. https://doi.org/10.1016/j.gloenvcha.2021.102354.

Shi, L., Lamb, Z., Qiu, X., Cai, H., and Vale, L. (2018). Promises and perils of collective land tenure in promoting urban resilience: Learning from China's urban villages. *Habitat International*, 77 (July), 1–11.

Spicer, J. S. (2020). Worker and community ownership as an economic development strategy: Innovative rebirth or tired retread of a failed idea?. *Economic Development Quarterly*, 34(4), 325–342.

Stephens, C. (1996). Healthy cities or unhealthy islands? The health and social implications of urban inequality. *Environment and Urbanization*, 8(2), 22.

Taylor, D. (2014). *Toxic Communities: Environmental Racism, Industrial Pollution, and Residential Mobility*. NYU Press.

Taylor, Z. J. (2020). The real estate risk fix: Residential insurance-linked securitization in the Florida metropolis. *Environment and Planning A: Economy and Space*, 55(6), 1131–1149. http://dx.doi.org/10.1177/0308518X19896579.

Teicher, H. M. (2018). Practices and pitfalls of competitive resilience: Urban adaptation as real estate firms turn climate risk to competitive advantage. *Urban Climate*, 25 (September), 9–21.

Tsing, A. L. (2005). *Friction: An Ethnography of Global Connection.* Princeton University Press.

United Nations. (2016). The World's Cities in 2016 [Moreno, E., Arimah, B., Otieno, R.O., Mbeche-Smith, U., Klen-Amin, A., Kamiya, M., et al.]. Statistical Papers—United Nations (Ser. A), Population and Vital Statistics Report. UN.

Van Sant, L. (2019). Land reform and the green new deal. *Dissent*, Fall. https://www.dissentmagazine.org/article/land-reform-and-the-green-new-deal.

Veracini, L. (2015). Introduction: The settler colonial present. In L. Veracini (Ed.), *The Settler Colonial Present* (pp. 1–12). Palgrave Macmillan.

von der Osten, F. B., Kirley, M., and Miller, T. (2017). Sustainability is possible despite greed—exploring the nexus between profitability and sustainability in common pool resource systems. *Scientific Reports*, 7(1), 2307. https://doi.org/10.1038/s41598-017-02151-y.

Winkler, J. (2016). The legacy of urban renewal in New York City. Parsons, the New School, New York City. https://due-parsons.github.io/methods3-fall2016/projects/the-legacy-of-urban-renewal-in-new-york-city-/.

CHAPTER 3

Budgeting for Climate Justice?

Contested Futures of Urban Finance

SARAH KNUTH

A growing consensus suggests that cities worldwide need a new generation of infrastructure to respond to the challenges of climate change. How these urban infrastructural transformations will be resourced and whose voices will be empowered to make these funding decisions remain open and imperative questions. There is growing mainstream interest in developing new financial markets for the low-carbon economy, climate change adaptation, and their infrastructures. Meanwhile, justice movements across many urban contexts are challenging inherited forms of urban austerity and financialization, including pervasive racialized exclusions and injustice in urban investment. This confrontation is provoking renewed questioning of the role of private finance versus governments and communities in driving urban development—with climate needs now a central vector of competing transformational programs.

In this chapter, I consider a "scalar solution" frequent within proposals to make cities worldwide more accommodating for mainstream climate finance—what Bigger and Webber (2021) term their would-be "reformatting" by actors like the World Bank in new programs of "green structural adjustment." In chapter 4 of this book, Silver further examines these urban climate experiments in action, discussing problematic experiences on the ground in sub-Saharan Africa. As Goldman (2005) has explored, cities of the majority world were functionally invisible to private international financiers in the twentieth century. In the name of unbearable risk, they generally refused loans to anyone but national governments (while frequently excluding them as well), and they required the mediation of the World Bank and other development finance institutions. Attempts to open up new frontiers in private sub-sovereign lending today seek novel pathways to rescale indebtedness, making more cities and communities visible to financial markets. In doing so, they frequently look to the United States as an influential rich-country model of decentralized urban finance: a relatively exceptional case in its massive, 150-year-old mu-nicipal bond market and particularly restless history of urban fiscal-financial experimentation.

I argue here that attempts to export the United States' decentralized model of urban finance, and constitutive tools such as revenue bonds and "value cap-ture" financial instruments (Weber 2010; Ashton, Doussard, and Weber 2012; Chen 2019), contain both a serious risk and a major paradox. Both, I suggest, threaten the animating goal of this book: to create just cities in an era of cli-mate change. First, the United States is a cautionary example of decentralized urban finance in practice, in its history of fiscal fragmentation, racialized ex-clusion, and private financial exploitation. Hyperlocalization and fragmenta-tion across many areas of U.S. public finance have been an enabling condition and force for race-class injustice. They have promoted undemocratic forms of fiscal control for some and starved others of resources. Moreover, they have created damaging relational geographies of "creative extraction," in which re-sources are actively siphoned out of Black places for investment in white places (Seamster and Purifoy 2021). U.S. cities are undoubtedly sites marked by "in-terlocking processes of uneven urban development and climate vulnerability," as argued in the introduction. Crucially, such processes of structural inequal-ity, racialized differentiation and austerity, and opportunistic extraction (Big-ger and Millington 2020; Ponder 2021; Seamster and Purifoy 2021) in how ur-ban governments secure financial resources partly explain why that is.

Second, an important reason why the U.S. model has fostered financial ex-perimentation is its frequently contentious politics and internal tensions. Suc-cessive waves of U.S. urban financial "innovation" have been in part urban governments and financiers' strategies to circumvent a history of restrictions placed on cities' spending and ability to take out more conventional forms of debt (Sbragia 1996; Davidson et al. 2017). Frequently, these austerity mea-sures were ushered in amid financial crises (often after speculative urban eco-nomic bubbles) or other political upheavals—for example, the populist "tax revolt" of the late 1970s and early 1980s that swept Reagan and neoliberalism into national power, and the Tea Party movement following the subprime cri-sis (Sbragia 1983, 1996; Peck 2012). U.S. influence in fiscal-financial policy cir-culations stems not only from its power as an international financial center but also from this fractious boom-bust legacy, and convoluted and problem-atically creative modes of public revenue generation urban governments have eked out within it. Many of these fiscal-financial experiments have proven di-sastrous; for example, in a wave of urban bankruptcies following the subprime crisis (e.g., Kirkpatrick and Smith 2011; Peck 2012; Ponder and Omstedt 2019).

Yet despite their deepening problems at home, these models are being actively exported to new places and sold as novel climate solutions.

I conclude by considering rising urban justice movements and their proposals for transforming these problematic urban financing practices—vital programs in advancing climate justice in and beyond U.S. cities. As part of their organizing for more economically and racially equitable cities and reparative modes of urbanism, movements such as Black Lives Matter and Right to the City frequently demand more local self-determination in public spending and sometimes direct democratic participation in budgeting. Movements such as prison abolition and Defund the Police are inspiring emancipatory visions for climate change adaptation (Ranganathan and Bratman 2019; Anguelovski et al. 2021; Purdum et al. 2021) to replace technocratic, exclusionary norms in mainstream resilience and disaster planning (Hardy, Milligan, and Heynen 2017; Jacobs 2019; Grove, Barnett, and Cox 2020).

However, participatory budgeting and other decentralized fiscal models face challenges in just climate response. In the United States, they inherit a challenging historical legacy, notably in Black-majority cities—for example, of progressive Black-led urban administrations in the later twentieth century winning more control over fewer resources, with budgets slashed by decades of suburbanization and fiscal fragmentation (as, for example, Self [2005] chronicles in Oakland, California). Furthermore, progressive visions for fiscal decentralization risk opportunistic capture in a financialized world. The large-scale resource demands of climate response underline the need for collective, actively redistributive interventions in both a social and geographic sense— as Bigger and Millington (2020) advocate, "rescaling borrowing to higher political scales that can more progressively distribute risks" (601). Recent movement calls for a Black climate mandate and Red, Black, and Green New Deal (Gulf Coast Center for Law and Policy and M4BL 2021) suggest pathways toward building these more expansive investment models, driving public resources at scale toward just urban climate outcomes.

Financialized Experimentation for the "Climate-Proof" City

Urban climate scholars like Bulkeley and Castán Broto (2013) have long explored the experimental power that cities and urban governments may exert in evolving solutions to climate change's challenges, a proposition with overarching significance for the programs discussed here (see also Rice 2010; Long and Rice 2019). However, in recent years that urban climate experimentation has taken a financialized turn. Making cities low-carbon and resilient means fundamentally reworking existing urban geographies and modes of city building, radical changes that require major upfront investments. Today, powerful institutions like the World Bank advance language of a "financing gap" to frame this fiscal challenge as a task for experimental finance and the creation of new forms of public and private indebtedness (Knuth 2015; Bigger and Millington 2020; Hilbrandt and Grubbauer 2020; Bigger and Webber 2021). For more than a decade, institutions like the World Bank and the OECD have also promoted a range of financial instruments in response to this call, both to urban governments and to institutional investors and banks seeking new investment frontiers (for example, World Bank 2010; Della Croce, Kaminker, and Stewart 2011; Merk et al. 2012).

The term "climate finance" was popularized in ongoing United Nations climate negotiations amid justice calls for high-emitting northern governments to fund climate change adaptation in developing countries (see, e.g., Knuth and Krishnan 2021). Though urban resilience remains central to World Bank experimentation (Bigger and Webber 2021), I frame the concept more expansively in this chapter to include a broad range of "low-carbon" investments also gaining speed internationally (see also Bridge et al. 2020). Across this broad investment field, discursive emphasis on climate-related financial need has been met with an evident opportunity in financial institutions' long struggle for yield since the subprime collapse and willingness to consider novel climate-facing investments in response (see, e.g., Johnson 2014; Knuth 2015, 2021a).

Proposed urban financing models for decarbonization and adaptation have ranged widely: from green municipal bonds to infrastructure trusts, schemes to link urban finance to carbon markets, value capture tools targeting new low-carbon or "green" value streams, and more. Relevant value streams include, for example, those prospectively generated from energy efficiency and "climate-proofing" retrofits to existing homes or commercial buildings. Urban governments and private financial partners may package, securitize, and sell this value to investors via new financial instruments like Property Assessed Clean Energy (PACE) and PACE bonds (Knuth 2016, 2019).

This wave of green experimentation is entangled with preexisting forms of urban infrastructural financialization and rent extraction. For example, major practices include the use of instruments such as project finance for renewable energy development and other low-carbon investment schemes (Langley 2018; Baker 2021; Knuth 2021b), joined by newer "assetization" models like securitization, Yieldcos, and green bonds (Bridge et al. 2020; see also Knuth 2018). For climate change adaptation, one important practice is financialized infrastructural privatization around drought and water provisioning (Bigger and Millington 2020). Models range from boutique, extractive bond structures for experimental green infrastructures such as ecological stormwater management (Christophers 2018) to schemes for using large infrastructure projects as engines for generating and stacking investor returns—including for similarly climate-relevant purposes like major water management and flood control systems (Loftus and March 2016; Colven 2017; Grafe and Hilbrandt 2019).

These schemes are joined by a broader set of financialized experiments that attempt to keep cities and urban property insurable against new climate threats—place-based crises attracting growing speculative interest in global (re)insurance markets and financial centers (Johnson 2013, 2014; Taylor 2020; Taylor and Weinkle 2020). In the United States, these private experiments are particularly significant as existing public models, like the National Flood Insurance Program, are threatened by rising risk premiums and deeper inherited tensions (Elliott 2019), while cities highly exposed to climate risks face combined crises of real estate devaluation and collapsed fiscal capacity. In part, this looming climate-related fiscal crisis comes from threatened municipal bond downgrades, making loans less accessible and affordable for urban governments (Chung 2019; Cox 2021)—private "creditworthiness" judgments that ignore structurally uneven fiscal legacies of racialized property dispossession and disproportionately target Black-majority cities (Ranganathan 2016; Phinney 2018; Ponder 2021). Digging deeper on questions of scale, it also speaks to the decentralized way that U.S. urban governments assemble their resources from the ground up, centrally relying on locally generated property taxes, land value capture tools, and related "urban growth machine" logics (Logan and Molotch 2007) in their revenue generation (Shi and Varuzzo 2020; Taylor 2020).

In crucial ways, today's pathways for climate-related financialization have been enabled by renewed pushes for fiscal decentralization, devolution, and pathways for direct private lending to cities. As histories of development finance have chronicled (e.g., Goldman 2005; Alami 2021), northern financial centers have a long history of bipolarity when it comes to investing in the Global South, torn between conflicting impulses. On the one hand, financiers have sought to profit from speculative booms in resource peripheries and frontiers of infrastructure development, in the *longue durée* of colonial and neocolonial carbon capitalism that Silver unpacks in chapter 4 of this book. On the other hand, during much of the twentieth century, these investors retreated from the risks of financial globalization, following late nineteenth-century financial crises in the peripheries of high imperialism (Alami 2021).

The history of the World Bank and other post–World War II development finance was shaped by private banks' protracted redlining of entire geographic regions, as investment in multilateral institutions' bonds came to stand in for creditworthiness judgments that banks were unwilling to take on themselves (Goldman 2005). A meaningful outcome of this history of financial exclusion has been the absence of sub-sovereign debt markets in most of the world: global financial institutions have primarily "seen" public debt at the national level. The new climate-related financial experiments discussed are increasingly working to change that, as private financial institutions and a more "financialized" World Bank (Gabor 2019) seek more direct and unmediated ways of investing in new products, sectors, and market territories—today often promoted through discourses of pro-poor "financial inclusion."

As Bigger and Webber (2021) have recently argued, this new sub-sovereign financial expansion risks fresh rounds of creditor-centered political disciplining, what they term green structural adjustment enacted at the urban level. Scholars are advancing multiple critiques of these urban climate finance experiments. Bigger and Webber situate today's activity in longer histories of structural adjustment and the broader financialization of the international development apparatus, building on interventions such as Mitchell and Sparke's (2016) "New Washington Consensus" and Gabor's (2019) "Wall Street Consensus." Postcolonial urban critiques (Robin and Castán Broto 2021; Robin 2021) seek to decenter understandings of climate finance developed from tacit northern norms, calling for more situated exploration of what actually existing climate interventions and relevant financing practices can look like in cities of the majority world. In chapter 4 (this book), Silver furthers a decolonial project, advancing an explicitly "anti-colonial understanding of the climate debt . . . [as] the primary mechanism through which a more just climate urbanism may be configured." This approach opens up new pathways for resisting and reversing today's wave of green structural adjustment, joining growing conversations on North-South ecological debt, climate reparations, and abolition ecologies (Táíwò and Cibralic 2020; Aronoff et al. 2021; Heynen and Ybarra 2021).

This chapter argues that it is also necessary to situate and critique the models that the new urban climate finance/financialization is taking up. Mainstream development literature often frames the U.S. instruments and infrastructures discussed in this chapter as "best practices," elements of a virtuous one-way north-to-south flow of knowledge and expertise. Majority world cities are encouraged to reshape themselves after such U.S. and northern urban models: "reformat[ting] local governments across the Global South into reflections of Northern counterparts that have the capacity to borrow on international credit markets" (Bigger and Webber 2021, 37). Whether or not these ambitious transformations are really plausible (Bigger and Millington [2020] argue that green bond uptake so far has been more superficial), the models themselves demand critical scrutiny.

A decolonial rethinking suggests reframing many of these northern urban finance models as cautionary tales. In the U.S. case, there is a strong argument for doing so even before the rollout of neoliberalism in the later twentieth century deepened urban governmental trends toward austerity, privatization, and financialization (Peck 2012). This reframing is particularly crucial as critiques break down distinctions between embedded traditions of financial imperialism "abroad" and racial liberalism and settler colonialism "at home" in U.S. cities. Historically and as reproduced in new ways today, "racial finance capitalism" (Ranganathan 2020), its constituent instruments, and their geographic circulations embed practices of racialized differentiation and relational extraction that are inextricably articulated and coproductive (see, e.g., Hall 2021 [1980]; Ranganathan 2016; Cowen 2020; Ponder 2021; Seamster and Purifoy 2021). These critiques are increasingly being advanced in practice through transformative movements like Black Lives Matter.

U.S. Fiscal Federalism: Evolving Instruments in a Decentralized System

As discussed, a notable trend in mainstream urban climate finance initiatives today is the export of U.S.-derived instruments like land value capture tools and revenue bonds, which originated in the country's fragmented political geographies and massive, historically experimental municipal bond market. New climate finance initiatives build on existing efforts to export and internationalize these models (see, e.g., Strickland 2013; Chen 2019; see also Weber 2010; Ashton, Doussard, and Weber 2012). Such schemes attempt to transplant and adapt these financial instruments to contexts, frequently a challenging prospect. Difficulties extend beyond the complications of climate-related financial inclusion, "greenlining" of cities and communities that major financiers long judged too risky or unprofitable for lending. Other obstacles come from the way many national states organize public finance and delegate cities' independent financial powers (or lack thereof). More centralized governments have been less structurally open to autonomous urban experiments, and they now face obstacles in adapting financial instruments grown in very different fiscal and financial geographies (Tapp and Kay 2019).

In contrast to more centralized states, the United States has been an unusually open space for fiscal financialization, present and past. Its decentralized federalist structure delegates many budgetary powers to (regional) states, who in turn have chosen to further devolve control to cities and other local authorities—sometimes hyperlocal ones, like special districts. This experience stands in contrast to more normal urban experiences worldwide. With private financiers willing to lend to highly local borrowing units (alongside a raft of tax incentives and intragovernmental subsidies for municipal bonds [Sbragia 1996]), this governmental decentralization has generated a massive municipal bond market. Over the last century and a half, that market has swelled to include the independent borrowing of a host of fragmented public (and public-private) institutions: public authorities, special districts, school districts, and more.

These fragmented governing units use a growing range of financial instruments, but most have been extrapolated from the revenue bond model. Revenue bonds' key, novel proposition in the mid-twentieth century was that infrastructure development could and should be made to "self-finance" or "pay for itself"—an idea that laid the groundwork for deepening privatization in the later twentieth century. Revenue bonds are issued within municipal bond markets, just like more traditional general obligation bonds. However, instead of being backed by the full faith and credit of borrowing governments and repaid to creditors out of constituents' taxes, they collect income from user fees, tolls, and other streams of income generated by infrastructure projects themselves. One particularly influential value capture model in U.S. cities today, tax increment finance (TIF), is a latter-day variant and elaboration. Through TIF districts and TIF bonds floated in municipal bond markets, urban governments partially recapture growth in land values and property taxes after redevelopment interventions, channeling these revenues back to city or redevelopment authority budgets (Weber 2010; Ashton, Doussard, and Weber 2012).

Revenue bonds were pioneered by Robert Moses in New York City in the mid-twentieth century, in conjunction with the related innovation of the public authority form (Caro 1974). Public authorities like redevelopment authorities are quasi-private entities embedded within the U.S. urban-regional state, which are empowered to self-direct themselves and recycle their revenue streams toward further development. Moreover, they are deliberately siloed from the obligation to seek voter approval for their decisions and spending—an achievement that took substantial legal finagling and remains dubious. This insulation from the demands of urban democratic participation and accountability to voters has for decades helped make them a favored tool in U.S. cities' economic (re)development spending.

A Checkered History: Privatism and Racialized Exclusion

Like so many other areas of U.S. politics, these financing practices have been indelibly shaped by racialized inequalities. The bitter history of racial capitalism in the United States (after Robinson 2020; see also Brahinsky 2014; Dillon 2014; Pulido 2017; Ponder and Omstedt 2019; and many others) has exerted particular force here. As historians of U.S. public finance have theorized (McDonald 1986; Einhorn 2001), the sectarian rifts of the Civil War helped produce a distinctive "privatism" in contemporary structures and instruments of U.S. public finance. This privatism is an important fiscal philosophical expression of U.S. racial liberalism and settler colonialism—one that has required structural amnesia about these regimes' roots in violent governmental property appropriations and redistribution (see also Blomley 2003; Ranganathan 2016, 2020). Privatism is characterized by hostility to collective action and overt government wealth redistribution, including geographic centralization and redistribution of wealth between places and regions. Its narrow scalar-political philosophy has recurrently cropped up since in grassroots cultural-political forms, for example, in populist antitax uprisings like the tax revolt of the late 1970s–early 1980s and the Tea Party movement of the 2010s (Sbragia 1983; Einhorn 2008; Peck 2012).

As Einhorn and McDonald explore, the philosophy of privatism was earlier crystallized within nineteenth-century U.S. fiscal instruments like special districts. These instruments and their embedded political philosophies are still very much alive in the disparate, ever-accumulating arsenal of U.S. public finance. For example, "Mello-Roos" special districts and associated instruments like "dirt bonds" have frequently been used to finance the construction of localized infrastructural systems like sewers, particularly in new suburbs (see, e.g., Fulton 1999). More recently, they have been used as a model for new low-carbon financial instruments like PACE bonds, intended to finance urban energy efficiency retrofits (Knuth 2016, 2019). In one sense, these processes are successful examples of direct democracy in action, since creating a special district typically requires a direct vote from property holders. However, as Einhorn (2001) has discussed, the directness of this tool simultaneously makes it highly exclusionary: it restricts decision-making, infrastructures, and their benefits to a narrowly defined community of interest, typically private property owners who directly benefit from infrastructure improvements. In many ways, that exclusion has been spatialized, racialized, and classed as well.

More broadly, this move to withdraw from more complex urban challenges and diverse polities, and the white racialized privilege of this withdrawal, has been a frequent refrain in critiques of U.S. suburbanization. A crucial insight into how U.S. inner cities were disinvested from the mid-twentieth century lies in analyses of how these waves of spatial decentralization were simultaneously moments of fiscal fragmentation. White suburban flight withdrew property tax resources from urban services and the maintenance of urban infrastructures, often already older and increasingly deteriorated (Self 2005; Sugrue 2014). Exclusionary white suburbs were free to build new greenfield infrastructures and to recirculate their already unequal property tax wealth locally—again, frequently via special districts and other community-led resourcing models. By jumping jurisdictional boundaries and forming new towns (and even cities and regions, as Sunbelt cities escalated this strategy to the next level), suburbs still successfully withhold tax dollars from the cities they surround. Frequently, this has meant starving needed resources from these places and associated communities of color—often just as progressive Black-led urban administrations and racial justice coalitions in Black-majority cities got a foot in the door politically, as Self (2005) explores in Oakland (see also Ponder 2021). These threads begin to suggest how the fragmented system in the United States has simultaneously allowed space for fiscal-financial experimentation and reproduced racialized exclusion.

The Opportunities and Dangers of Financial "Partners"

In crucial ways, these highly complex U.S. fiscal geographies reflect frequent financial crises—and recurrent moments of postcrisis austerity and fiscal disciplining—in U.S. urban history. State governments may choose to delimit or retract cities' powers to spend and to take on debt, and they have historically exercised this authority in the wake of crises—adding public disciplinary power to the private financial differentiation and gatekeeping role of credit ratings agencies (Sbragia 1983; Hackworth 2007). Sbragia (1996) and Weiss (1991) track this history of urban-national crises, with post-bust city administrations recurrently facing large debt repayment burdens and an outraged citizenry (see also McDonald 1986; Scobey 2002). Limitations on taxation, spending, and debt powers have also come via mass popular referendums in certain states, for example, the tax revolt case of Proposition 13 in California. California cities have experienced decades of fiscal struggles and new financialized experimentation in the wake of Prop 13's limits on conventional property taxation (Sbragia 1983; Kirkpatrick and Smith 2011). Frequently, these governmental crackdowns have been loaded with anti-urban, anti-immigrant, and racist overtones.

At the same time, this turbulence has helped drive generations of fiscal-financial experiments, often in times when cities' power has been foreclosed from above (by state governments) or below (in voters' rejections of new proposed taxes or bond measures). Sbragia (1996) has argued that rather than legally overturning these constraints, cities seeking renewed growth have typically circumvented them via the creation of new fiscal-financial instruments—for example, forms of public indebtedness that are not (yet) formally classified as debt. Notably, these experiments have also been enabled by the settler-colonial wealth and exceptional power of the United States as a financial center. This positioning has presented available capital, markets for floating new kinds of municipal bonds, and willing financial industry partners drawn by the lure of new accumulation opportunities.

These accumulated loopholes and work-arounds have allowed U.S. cities to legally but surreptitiously assume new debt burdens in off-the-books "shadow" debt (Pacewicz 2013). They have also required partnerships with the financial industry and actors capable of creating and issuing new types of municipal bonds. Who, precisely, has driven these experiments is a matter of interpretation, particularly as historical-geographical shifts and place-based variation in governmental versus private financial power mean that answers vary. Certainly, however, cities operate under a degree of compulsion. Critical scholars have underlined how the neoliberal withdrawal of New Deal–Keynesian federal funds to cities heightened urban governments' dependence on private capital and intensified their competition for footloose investment—made so in part by parallel moves for financial deregulation (Harvey 1989; Leitner 1990).

U.S. cities' reliance on exotic fiscal instruments and off-the-books borrowing practices came under increased scrutiny amid the subprime crisis. In the collapse's interconnected urban fiscal crisis (Peck 2012), the once-secure U.S. municipal bond market was severely strained. As noted earlier, the years after the crisis saw a swathe of high-profile defaults and bankruptcies, including Detroit, several California cities, and Harrisburg, Pennsylvania (Kirkpatrick and Smith 2011)—again with highly racialized dynamics and structural inequalities for Black-majority cities (Phinney 2018; Ponder and Omstedt 2019). Whole segments of the U.S. municipal bond market were permanently taken out by the subprime crisis, while complex "conduit" value capture instruments like TIF bonds experienced an especially high failure rate (SEC 2012). Meanwhile, many cities suffered the consequences of bad fiscal-financial deals contracted during the subprime bubble. Financial players searching for higher yields marketed more exotic and higher-return fiscal instruments to urban and government agencies. Notably, many cities took on ultimately disastrous interest-rate swaps (IRSs), boutique financial arrangements marketed by investment banks as useful for lowering the costs of finance in a boom. When that boom collapsed immediately thereafter, the terms of IRS agreements nonetheless stuck cities with paying exorbitantly high interest rates to investment banks, or sums ranging in the millions to buy their way out of these deals. This experience provoked widespread protest, including the Occupy movement.

A recurring theme in this experience is unequal power between U.S. urban governments using these tools and the financial industry, to which they have turned as a problematic ally. Contemporary investigations from entities such as the Pennsylvania Budget and Policy Center (Ward 2012), Refund Transit Coalition (2012), and State Budget Crisis Task Force (2012) exposed how financiers actively marketed exotic IRSs to relatively unsophisticated borrowers like school districts, including in historically disinvested regions like the Rust Belt (see also Ponder and Omstedt 2019). As with subprime loans, banks pitched their expertise and new financial innovations to manage risk, while urban governments took on these instruments without adequately understanding them. These patterns suggest that the decentralized U.S. fiscal system is worryingly able to offer up vulnerable client governments for predatory financiers. Moreover, shadow debt systems make the total amount and nature of U.S. urban debt highly difficult to ascertain, since so much U.S. local debt is now effectively off the books, not assessed in compiling municipalities' credit ratings, and not subject to popular vote. Collectively, these localist financialized experiments have produced a system that is frequently complex, opaque, and impenetrable when viewed in aggregate, hostile in many ways to meaningful democratic decision-making.

Urban Climate Justice Movements Rethink the Budget

In the contemporary moment, movements for racial justice have risen across the United States with calls to defund the police and abolish the country's prison-industrial complex (see, for example, Gilmore 2007; Loyd, Mitchelson, and Burridge 2012; Camp and Heatherton 2016). These broad-based mobilizations have called out the top-heavy public resources that U.S. cities dedicate to racialized policing of Black and Brown neighborhoods, and to bolstering the imagined security of private property and private property owners, a class that has always skewed white in the United States and grown whiter since the subprime crisis. Equally, they draw sharp attention to resources not spent in heavily policed places: on schools, on repairing crumbling infrastructure, on reinvesting in public housing, on good jobs.

Demands for divesting from police forces and their increasingly military-grade hardware are long-standing among U.S. urban justice movements. Equally, so are calls for reallocating these funds to resource-starved social institutions, in places that have frequently become ground zero for new fiscal-financial predations and "creative extraction" (Seamster and Purifoy 2021) on top of systematic race-class exclusions. However, these efforts are building new levels of organization and political traction today, particularly as Black Lives Matter mobilizations building from 2013 increasingly translate movement energy into policy. For example, dozens of Black-led organizations are now collaborating under the banner of Movement for Black Lives (M4BL) and crystallizing movement investment and divestment demands in their national policy platform: "We want investments in Black communities, determined by Black communities, and divestment from exploitative forces including prisons, fossil fuels, police, surveillance and exploitative corporations" (M4BL 2021a).

These mobilizations open a window onto the terrain where movements for urban racial justice and climate justice meet. New initiatives suggest the future of this conjoined organizing moving into an uncertain twenty-first century. For example, M4BL's policy platform (2021a) underlines divestment in the U.S. prison-industrial complex and its fossil fuel apparatus as an articulated project and calls out just low-carbon infrastructures as part of broader community investment needs: "a divestment from industrial multinational use of fossil fuels and investment in community-based sustainable energy solutions." Movement players are now advancing large-scale programs for racially just climate investment and divestment. The Gulf Coast Center for Law and Policy and M4BL (2021b) have formulated the National Black Climate Mandate and call for a Red, Black, and Green New Deal, which they are now building into a raft of concrete policy proposals.

These movement initiatives are developing in conversation with critical climate scholarship. Recent interventions flesh out the philosophical and concrete relations articulating racial and climate justice projects, for example, programs of prison abolition and disaster risk mitigation under climate change (Purdum et al. 2021) and fossil fuel divestment and climate reparations (Táíwò and Cibralic 2020; Aronoff et al. 2021). These arguments emphasize fossil extraction regimes' legacy harms to frontline communities and underline the need for transitions justice within broader fossil fuel divestment campaigns (see, e.g., Knuth 2017). Today's mobilizations increasingly draw out the many ways in which urban justice problems are climate justice problems, and always have been. For example, the racialized processes like suburban sprawl surveyed here have helped bake fossil energy intensity and emissions into U.S. built environments while framing the relational wealth extraction that shapes much current U.S. climate vulnerability (Knuth 2010, 2019; see also Ponder 2021; Seamster and Purifoy 2021). Urban climate justice movements are helping drive these buried relations to the surface (Cohen 2017).

These contributions join growing scholarly challenges to mainstream models of climate change vulnerability and urban resilience planning, and visions for analytical and political alternatives (Jacobs 2019; Long and Rice 2019). Critics have tracked uses and abuses of notions of resilience by neoliberal urban administrations (Derickson 2016), including maladaptive outcomes like new forms of gentrification, displacement, and compounded vulnerability. Also influenced by movements like Black Lives Matter, scholars and activists increasingly link these critiques to calls for more transformational abolitionist and emancipatory programs (Ranganathan and Bratman 2019; Anguelovski et al. 2021; Heynen and Ybarra 2021).

More particularly, racial justice calls for community self-determination (M4BL 2021a) echo many of the critiques raised earlier of the U.S. fiscal-financial model, particularly its constitutive racialized exclusions. Too often, questions of taxation, public finance, and budgeting are still presented as an essentially "post-political" arena of governance: opaque, technically complex, boring—a separate sphere for qualified experts and objective rationality above the fray of political dispute, and certainly the emotive demands of mass mobilizations. Movements like prison abolition and Defund the Police underline such characterizations as the ideologically loaded political constructions they are. How public monies are obtained and spent, on what, for whom, and who is empowered to make such calls are all essentially political decisions—particularly where major segments of the population have been systematically excluded from state resources, exposed to injustice through such spending, and otherwise have a claim on public reparations.

Self-determination is an important conceptual plank within broader discussions of racial reparations (Táíwò 2021). Translated to urban policy, the concept recurs in calls like M4BL's (2021a) invest-divest platform and in demands for Black community control over investment. These arguments suggest that taking more control over budget making can help marginalized communities give voice to injustices past and present, demand recognition from the state, win needed resources, and shape how those resources should be spent (or what they should be withdrawn from). Hardy, Milligan, and Heynen (2017) and Jacobs (2019) discuss how these ideas can translate into more just and effective climate adaptation and disaster mitigation planning.

As racial climate justice movements increasingly rethink urban governance and public budgets, they may make common cause with broader urban organizing, notably the international Right to the City mobilizations that inspire this book. For example, in centering community self-determination as a pathway toward just climate futures, M4BL's (2021b) Red, Black, and Green New Deal vision frequently resonates with Right to the City's rights and liberties-based framework: "We believe all Black people have the right to determine our own futures; where we can earn a decent living, purchase a home, raise a family and live in a safe community with access to reliable, clean and affordable services. We should be able to add our voice to the problems and solutions that impact our family and our community's well being. We should be able to build a future around the things that are important to us, leaving a legacy of generational and cultural value for those that come after us."

It is useful to further consider these calls for Black self-determination (in the United States and beyond) within broader international movements to restructure and democratize public spending. These mobilizations include the circulations of political discourses like the Right to the City, but also ways in which they have been operationalized in concrete policy—for example, direct democratic tools like urban participatory budgeting and governmental restructuring ideas such as fiscal devolution. Strikingly, these movements frequently offer their own version of the decentralized "scalar solution" now being take up by mainstream climate finance and its institutional advocates. Reorganizations like devolution and participatory budgeting similarly centralize rescaling as a pathway to ceding local places more control over public spending: matching locally felt needs with more localized control over resources. Like U.S. calls for defunding police, these moves have frequently been framed as a strategy for redirecting resources—and better kinds of resources—toward communities historically marginalized and underserved by more centralized, autocratic, or structurally unjust states (Baiocchi and Ganuza 2014).

However, the cautionary localist experience centered in this chapter suggests a need for care on how programs of self-determination are organized and funded, and at what scale. Such initiatives walk a tightrope in gathering needed resources for climate action. They must avoid pressures on already burdened households and communities, while matching the scale of investment needed for adaptation or decarbonization—costs that may be overwhelming for highly exposed cities and communities. Meanwhile, they risk fresh city-by-city and community-by-community unevenness and co-optation, as well as financial exploitation amid growing investor interest in climate change. Existing experience suggests other cautionary tales of localist fiscal experiments and climate self-determination efforts gone wrong, for example, some progressive participatory budgeting initiatives that have lost their political teeth in becoming neoliberal "good governance" models (Baiocchi and Ganuza 2014) or shallower "participation" initiatives in local climate planning (Grove, Barnett, and Cox 2020). Other participatory budgeting efforts have been captured by real estate developers, financiers, and programs of real estate value-led development (Pimentel Walker 2015).

The U.S. experience surveyed here further argues against an overly simple equation of devolved budgetary control with just outcomes, and suggests many ways that private financial capital can exploit fragmented local governments. Decentralization and even instances of successful direct democracy (e.g., suburban special districts) within U.S. public finance have often entrenched racially unequal wealth and resource exclusion rather than constituting a liberatory alternative. In this context, movements like M4BL make a useful intervention in linking community self-determination over funding decisions to more expansive and redistributive resourcing visions. M4BLs (2021a) invest-divest vision is explicitly multiscalar, advancing concrete calls for "reallocation of funds at the federal, state and local level." Similarly, movement take-up of the Green New Deal framework suggests a large-scale vision of social and geographic redistribution. Visions like a Red, Black, and Green New Deal create opportunities to transformatively reimagine urban climate finance for the United States (and potentially beyond), as organizers seek new tools and frameworks for resourcing community-determined climate needs.

Conclusion

Contemporary justice mobilizations pose important challenges to the U.S. urban financing model critiqued here, exposing the country's racially and economically unjust fiscal histories and envisioning new investment and divestment practices for decarbonization and climate change adaptation. The same legacy landscapes of suburbanization, privatism, and urban financial exclusion that help drive U.S. greenhouse emissions have also cultivated harmful forms of public finance, indebtedness, and financialized exploitation. On balance, the U.S. case suggests a cautionary tale for initiatives that seek governmental decentralization and localism as an easily applied "scalar solution" in resourcing climate response. That is true for both mainstream climate investment discourses that seek to export the U.S. model as a "best practice" for cities of the majority world and would-be progressive alternatives like participatory budgeting. While cities' climate investment needs will be diverse and may require a diverse resourcing toolkit, all tools come with embedded philosophies, costs, and trade-offs that must be chosen with eyes open.

In important ways, this critique reinforces the more centralized and collective vision suggested by a Green New Deal (Aronoff et al. 2019; see also Elliott 2019; Bigger and Webber 2021), particularly new visions developing in and through the Black Lives Matter movement (Gulf Coast Center for Law and Policy and M4BL 2021). A localist fiscal-financial tool like the special district may succeed in promoting direct democratic participation, particularly in homogeneous communities who see its use as enlightened self-interest. Conversely, it is highly limited when faced with broader collective needs and challenges, more expensive and massive infrastructural systems, and any project that seeks to redistribute wealth socially or geographically. It is crucial to match governmental resources and tools, including ones for wealth redistribution, to the large-scale, costly, and spatially complex challenges of just climate response. Despite the original New Deal's role in creating the tough climate problems that the United States and world face today, it remains notable in U.S. history as then-unprecedented governmental restructuring to make the country work in more centralized, coordinated, and public welfare–oriented ways. The original New Deal both substantially disciplined the financial sector and created many tools for mobilizing public resources, redistributing these funds to cities and otherwise assembling resources at scale.

U.S. mobilizations for a more progressive and racially just Green New Deal must ambitiously exploit the possibilities of such a collective resourcing program, a badly needed alternative to problematic U.S. fiscal-financial models now seeing similarly problematic dissemination abroad. Simultaneously, organizers must strenuously build and support pathways for holding a more centralized redistributive program accountable to frontline communities most exposed to climate harms. A more radically just Green New Deal must support long-marginalized communities' claims on reparative resources and provide actionable spaces for engaging them, again prioritizing communities chronically excluded from governmental resources. Fundamentally, this expansive systemic reimagination must raise the voices of marginalized cities and communities while mobilizing greater collective resources than they can command alone—a noteworthy departure for communities that have historically encountered the state as a locus of injustice and exclusion.

NOTES

This chapter was greatly improved by generative conversations with members of the Urban Climate Finance Research Network, with particular thanks to Emma Colven, Fritz-Julius Grafe, Hanna Hilbrandt, Sage Ponder, Enora Robin, and Zac Taylor.

REFERENCES

Alami, I. (2021). Global finance capital and third world debt. In I. Ness and Z. Cope (Eds.), *The Palgrave Encyclopedia of Imperialism and Anti-Imperialism*, 2nd ed. Palgrave Macmillan.

Anguelovski, I., Brand, A. L., Ranganathan, M., and Hyra, D. (2021). Decolonizing the green city: From environmental privilege to emancipatory green justice. *Environmental Justice*, OnlineFirst. https://doi.org/10.1089/env.2021.0014.

Aronoff, K., Battistoni, A., Cohen, D. A., and Riofrancos, T. (2019). *A Planet to Win: Why We Need a Green New Deal*. Verso.

Aronoff, K., Kozul-Wright, R., Rehman, A., Riofrancos, T., and Táíwò, O. O. (2021). A reparative politics for the climate crisis: A roundtable. *Dissent*, 68(2), 66–77.

Ashton, P., Doussard, M., and Weber, R. (2012). The financial engineering of infrastructure privatization: What are public assets worth to private investors? *Journal of the American Planning Association*, 78(3), 300–312.

Baiocchi, G., and Ganuza, E. (2014). Participatory budgeting as if emancipation mattered. *Politics and Society*, 42(1), 29–50.

Baker, L. (2021). Procurement, finance and the energy transition: Between global processes and territorial realities. *Environment and Planning E: Nature and Space*, OnlineFirst. https://doi.org/10.1177/2514848621991121.

Bigger, P., and Millington, N. (2020). Getting soaked? Climate crisis, adaptation finance, and racialized austerity. *Environment and Planning E: Nature and Space*, 3(3), 601–623.

Bigger, P., and Webber, S. (2021). Green structural adjustment in the World Bank's Resilient City. *Annals of the American Association of Geographers*, 111(1), 36–51.

Blomley, N. (2003). Law, property, and the geography of violence: The frontier, the survey, and the grid. *Annals of the Association of American Geographers*, 93(1), 121–141.

Brahinsky, R. (2014). Race and the making of southeast San Francisco: Towards a theory of race-class. *Antipode*, 46(5), 1258–1276.

Bridge, G., Bulkeley, H., Langley, P., and van Veelen, B. (2020). Pluralizing and problematizing carbon finance. *Progress in Human Geography*, 44(4), 724–742.

Bulkeley, H., and Castán Broto, V. (2013). Government by experiment? Global cities and the governing of climate change. *Transactions of the Institute of British Geographers*, 38(3), 361–375.

Camp, J., and Heatherton, C. (Eds.). (2016). *Policing the Planet: Why the Policing Crisis Led to Black Lives Matter*. Verso.

Caro, R. A. (1974). *The Power Broker: Robert Moses and the Fall of New York*. Knopf.

Chen, H.-Y. (2019). Cashing in on the sky: Financialization and urban air rights in the Taipei Metropolitan Area. *Regional Studies*, OnlineFirst. https://doi.org/10.1080/00343404.2019.1599104.

Christophers, B. (2018). Risk capital: Urban political ecology and entanglements of financial and environmental risk in Washington, D.C. *Environment and Planning E: Nature and Space*, 1(1–2), 144–164.

Chung, C. S. (2019). Rising tides and rearranging deckchairs: How climate change is reshaping infrastructure finance and threatening to sink municipal budgets. *Georgetown Environmental Law Review*, 32, 165–226.

Cohen, D. A. (2017). The other low-carbon protagonists: Poor people's movements and climate politics in São Paulo. In M. Greenberg and P. Lewis (Eds.), *The City Is the Factory: New Solidarities and Spatial Strategies in an Urban Age* (pp. 140–157). Cornell University Press.

Colven, E. (2017). Understanding the allure of big infrastructure: Jakarta's Great Garuda Sea Wall project. *Water Alternatives*, 10(2), 250–264.

Cowen, D. (2020). Following the infrastructures of empire: Notes on cities, settler colonialism, and method. *Urban Geography*, 41(4), 469–486.

Cox, S. (2021). Inscriptions of resilience: Bond ratings and the government of climate risk in Greater Miami, Florida. *Environment and Planning A: Economy and Space*, OnlineFirst. https://doi.org/10.1177/0308518X211054162.

Davidson, M., Ward, K., Jonas, A. E., Knuth, S., Weber, R., Wilson, D., and Sbragia, A. (2017). Book Review: *Debt Wish: Entrepreneurial Cities, U.S. Federalism, and Economic Development*. University of Pittsburgh Press, 1996. *Urban Geography*. http://dx.doi.org/10.1080/02723638.2017.1349990.

Della Croce, R., Kaminker, C., and Stewart, F. (2011). *The Role of Pension Funds in Financing Green Growth Initiatives*. OECD.

Derickson, K. D. (2016). Resilience is not enough. *City*, 20(1), 161–166.

Dillon, L. (2014). Race, waste, and space: Brownfield redevelopment and environmental justice at the Hunters Point Shipyard. *Antipode*, 46(5), 1205–1221.

Einhorn, R. L. (2001). *Property Rules: Political Economy in Chicago, 1833–1872*. University of Chicago Press.

———. (2008). *American Taxation, American Slavery*. University of Chicago Press.

Elliott, R. (2019). "Scarier than another storm": Values at risk in the mapping and insuring of U.S. floodplains. *British Journal of Sociology*, 70(3), 1067–1090.

Fulton, W. B. (1999). *Guide to California Planning*. Solano Press.

Gabor, D. (2019). *Securitization for Sustainability: Does It Help Achieve the Sustainable Development Goals?*. Heinrich Böll Stiftung.

Gilmore, R. W. (2007). *Golden Gulag: Prisons, Surplus, Crisis, and Opposition in Globalizing California*. University of California Press.

Goldman, M. (2005). *Imperial Nature: The World Bank and Struggles for Social Justice in the Age of Globalization*. Yale University Press.

Grafe, F. J., and Hilbrandt, H. (2019). The temporalities of financialization: Infrastructures, dominations and openings in the Thames Tideway Tunnel. *City*, 23(4–5), 606–618.

Grove, K., Barnett, A., and Cox, S. (2020). Designing justice? Race and the limits of recognition in greater Miami resilience planning. *Geoforum*, 117, 134–143.

Gulf Coast Center for Law and Policy and Movement for Black Lives (M4BL). (2021). National Black Climate Mandate. https://redblackgreennewdeal.org/wp-content/uploads/2021/10/Red-Black-Green-Climate-Mandate_NEWPRINT.pdf.

Hackworth, J. (2007). *The Neoliberal City: Governance, Ideology, and Development in American Urbanism*. Cornell University Press.

Hall, S. (2021 [1980]). Race, articulation, and societies structured in dominance. Chapter 11 in P. Gilroy and R. W. Gilmore (Eds.), *Selected Writings on Race and Difference* (pp. 195–245). Duke University Press.

Hardy, R. D., Milligan, R. A., and Heynen, N. (2017). Racial coastal formation: The environmental injustice of colorblind adaptation planning for sea-level rise. *Geoforum*, 87, 62–72.

Harvey, D. (1989). From managerialism to entrepreneurialism: The transformation in urban governance in late capitalism. *Geografiska Annaler: Series B, Human Geography*, 71(1), 3–17.

Heynen, N. and Ybarra, M. (2021). On abolition ecologies and making "freedom as a place." *Antipode*, 53(1), 21–35.

Hilbrandt, H., and Grubbauer, M. (2020). Standards and SSOs in the contested widening and deepening of financial markets: The arrival of green municipal bonds in Mexico City. *Environment and Planning A: Economy and Space*, 52(7), 1415–1433.

Jacobs, F. (2019). Black feminism and radical planning: New directions for disaster planning research. *Planning Theory*, 18(1), 24–39.

Johnson, L. (2013). Catastrophe bonds and financial risk: Securing capital and rule through contingency. *Geoforum*, 45, 30–40.

———. (2014). Geographies of securitized catastrophe risk and the implications of climate change. *Economic Geography*, 90(2), 155–185.

Kirkpatrick, L. O., and Smith, M. P. (2011). The infrastructural limits to growth: Rethinking the urban growth machine in times of fiscal crisis. *International Journal of Urban and Regional Research*, 35(3), 477–503.

Knuth, S. E. (2010). Addressing place in climate change mitigation: Reducing emissions in a suburban landscape. *Applied Geography*, 30(4), 518–531.

———. (2015). Global finance and the land grab: Mapping twenty-first century strategies. *Canadian Journal of Development Studies/Revue canadienne d'études du développement*, 36(2), 163–178.

———. (2016). Seeing green in San Francisco: City as resource frontier. *Antipode*, 48(3), 626–644.

———. (2017). Green devaluation: Disruption, divestment, and decommodification for a green economy. *Capitalism Nature Socialism*, 28(1), 98–117.

———. (2018). "Breakthroughs" for a green economy? Financialization and clean energy transition. *Energy Research and Social Science*, 41, 220–229.

———. (2019). Cities and planetary repair: The problem with climate retrofitting. *Environment and Planning A: Economy and Space*, 51(2), 487–504.

———. (2021a). Fictions of safety: Defensive storylines in global property investment.

Chapter 3 in D. A. Ghertner and R. W. Lake (Eds.), *Land Fictions: The Commodification of Land in City and Country* (pp. 62–85). Cornell University Press.

———. (2021b). Rentiers of the low-carbon economy: Renewable energy's extractive fiscal geographies. *Environment and Planning A: Economy and Space*, OnlineFirst. https://doi.org/10.1177/0308518X211062601.

Knuth, S., and Krishnan, A. (2021). British Academy COP26 Briefing: Climate Finance for Cities and Urban Governments. The British Academy, London.

Langley, P. (2018). Frontier financialization: Urban infrastructure in the United Kingdom. *Economic Anthropology*, 5(2), 172–184.

Leitner, H. (1990). Cities in pursuit of economic growth: The local state as entrepreneur. *Political Geography Quarterly*, 9(2), 146–170.

Loftus, A., and March, H. (2016). Financializing desalination: Rethinking the returns of big infrastructure. *International Journal of Urban and Regional Research*, 40(1), 46–61.

Logan, J. R., and Molotch, H. (2007). *Urban Fortunes: The Political Economy of Place*. University of California Press.

Long, J., and Rice, J. L. (2019). From sustainable urbanism to climate urbanism. *Urban Studies*, 56(5), 992–1008.

Loyd, J., Mitchelson, M., and Burridge, A. (2012). *Beyond Walls and Cages: Prisons, Borders, and Global Crisis*. University of Georgia Press.

McDonald, T. J. (1986). *The Parameters of Urban Fiscal Policy: Socioeconomic Change and Political Culture in San Francisco, 1860–1906*. University of California Press.

Merk, O., Saussier, S., Staropoli, C., Slack, E., and Kim, J.-H. (2012). Financing Green Urban Infrastructure. OECD Regional Development Working Papers 2012/10, OECD.

Mitchell, K., and Sparke, M. (2016). The new Washington consensus: Millennial philanthropy and the making of global market subjects. *Antipode*, 48(3), 724–749.

Movement for Black Lives (M4BL). (2021a). Invest-Divest. https://m4bl.org/policy-platforms/invest-divest/.

Movement for Black Lives (M4BL). (2021b). Red Black and Green New Deal. https://redblackgreennewdeal.org/.

Pacewicz, J. (2013). Tax increment financing, economic development professionals, and the financialization of urban politics. *Socio-Economic Review*, 11(3), 413–440.

Peck, J. (2012). Austerity urbanism: American cities under extreme economy. *City*, 16(6), 626–655.

Phinney, S. (2018). Detroit's municipal bankruptcy: Racialised geographies of austerity. *New Political Economy*, 23(5), 609–626.

Pimentel Walker, A. P. (2015). The conflation of participatory budgeting and public-private partnerships in Porto Alegre, Brazil: The construction of a working-class mall for street hawkers. *Economic Anthropology*, 2(1), 165–184.

Ponder, C. S. (2021). Spatializing the municipal bond market: Urban resilience under racial capitalism. *Annals of the American Association of Geographers*, OnlineFirst. https://doi.org/10.1080/24694452.2020.1866487.

Ponder, C. S., and Omstedt, M. (2019). The violence of municipal debt: From interest rate swaps to racialized harm in the Detroit water crisis. *Geoforum*, OnlineFirst. https://doi.org/10.1016/j.geoforum.2019.07.009.

Pulido, L. (2017). Geographies of race and ethnicity II: Environmental racism, racial capitalism and state-sanctioned violence. *Progress in Human Geography*, 41(4), 524–533.

Purdum, C., Henry, F., Rucker, S., Williams, D. A., Thomas, R., Dixon, B., and Jacobs, F. (2021). No justice, no resilience: Prison abolition as disaster mitigation in an era of climate change. *Environmental Justice*, OnlineFirst. https://doi.org/10.1089/env .2021.0020.

Ranganathan, M. (2016). Thinking with Flint: Racial liberalism and the roots of an American water tragedy. *Capitalism Nature Socialism*, 27(3), 17–33.

———. (2020). Empire's infrastructures: Racial finance capitalism and liberal necropolitics. *Urban Geography*, 41(4), 492–496.

Ranganathan, M., and Bratman, E. (2019). From urban resilience to abolitionist climate justice in Washington, D.C. *Antipode*, OnlineFirst. https://doi.org/10.1111/anti.12555.

Refund Transit Coalition. (2012). Riding the Gravy Train: How Wall Street Is Bankrupting Our Public Transit Agencies by Profiteering off of Toxic Swap Deals. June. https:// acrecampaigns.org/wp-content/uploads/2020/04/RidingtheGravyTrain-Jun2012.pdf.

Rice, J. L. (2010). Climate, carbon, and territory: Greenhouse gas mitigation in Seattle, Washington. *Annals of the Association of American Geographers*, 100(4), 929–937.

Robin, E. (2021). Rethinking the geographies of finance for urban climate action. *Transactions of the Institute of British Geographers*, OnlineFirst. https://doi.org/10.1111/tran .12508.

Robin, E., and Castán Broto, V. (2021). Towards a postcolonial perspective on climate urbanism. *International Journal of Urban and Regional Research*, 45(5), 869–878.

Robinson, C. J. (2020). *Black Marxism, Revised and Updated Third Edition: The Making of the Black Radical Tradition*. UNC Press.

Sbragia, A. (Ed.). (1983). *The Municipal Money Chase: The Politics of Local Government Finance*. Westview Press.

———. (1996). *Debt Wish: Entrepreneurial Cities, U.S. Federalism, and Economic Development*. University of Pittsburgh Press.

Scobey, D. M. (2002). *Empire City: The Making and Meaning of the New York City Landscape*. Temple University Press.

Seamster, L., and Purifoy, D. (2021). What is environmental racism for? Place-based harm and relational development. *Environmental Sociology*, 7(2), 110–121.

SEC. (2012). Report on the Municipal Securities Market. U.S. Securities and Exchange Commission, July 31.

Self, R. O. (2005). *American Babylon: Race and the Struggle for Postwar Oakland*. Princeton University Press.

Shi, L., and Varuzzo, A. M. (2020). Surging seas, rising fiscal stress: Exploring municipal fiscal vulnerability to climate change. *Cities*, 100, 102658.

State Budget Crisis Task Force. (2012). Report of the State Budget Crisis Task Force, July 2012.

Strickland, T. (2013). The financialisation of urban development: Tax increment financing in Newcastle upon Tyne. *Local Economy*, 28(4), 384–398.

Sugrue, T. J. (2014). *The Origins of the Urban Crisis: Race and Inequality in Postwar Detroit*. Rev. ed. Princeton University Press.

Táíwò, O. O. (2021). Reconsidering reparations. Chapter 4 in B. Hogan, M. Cholbi, A.

Madva, and B. S. Yost (Eds.), *The Movement for Black Lives: Philosophical Perspectives* (pp. 93–115). Oxford University Press.

Táíwò, O. O., and Cibralic, B. (2020). The case for climate reparations. *Foreign Policy*, 10 October. https://foreignpolicy.com/2020/10/10/case-for-climate-reparations-crisis-migration-refugees-inequality/.

Tapp, R., and Kay, K. (2019). Fiscal geographies: "Placing" taxation in urban geography. *Urban Geography*, 40(4), 573–581.

Taylor, Z. J. (2020). The real estate risk fix: Residential insurance-linked securitization in the Florida metropolis. *Environment and Planning A: Economy and Space*, 52(6), 1131–1149.

Taylor, Z. J., and Weinkle, J. L. (2020). The riskscapes of re/insurance. *Cambridge Journal of Regions, Economy and Society*, 13(2), 405–422.

Ward, S. (2012). Too Big to Trust? Banks, Schools, and the Ongoing Problem of Interest Rate Swaps. Pennsylvania Budget and Policy Center, Harrisburg. http://pennbpc.org/sites/pennbpc.org/files/TooBigSwaps.pdf.

Weber, R. (2002). Extracting value from the city: Neoliberalism and urban redevelopment. *Antipode*, 34(3), 519–540.

———. (2010). Selling city futures: The financialization of urban redevelopment policy. *Economic Geography*, 86(3), 251–274.

Weiss, M. A. (1991). The politics of real estate cycles. *Business and Economic History*, 20, 127–135.

World Bank (2010). A City-wide Approach to Carbon Finance. Carbon Partnership Facility Innovation Series, Carbon Finance Unit, World Bank, Washington, D.C.

CHAPTER 4

Climate Urbanism as Green Structural Adjustment

Unequal Center-Periphery Relations
in the Age of Climate Crisis

JONATHAN SILVER

This chapter situates climate urbanism within the longer histories of global, uneven center-periphery relations and how these are made visible at the urban scale in an age of climate crisis. Climate urbanism draws attention to a new planetary era of governing climate-changed cities. As Castán Broto, Robin, and While (2020) argue, "Climate change dramatically reshapes how we understand, imagine, live, and intervene in cities" (1). It is therefore vital to examine how climate urbanism is being configured, by whom, and through what operational logics. This is a scholarly and political imperative given that "we find ourselves in an intensified 'climate moment' for cities, as climate change transforms both how we live in urban areas and how we govern them in fundamental ways" (Castán Broto, Robin, and While 2020, 2). Central to this focus is exploring whether climate urbanism is perpetuating unequal, center-periphery relations through modes of neocolonial, technoenvironmental restructuring across sub-Saharan Africa.

To understand the shifting urban geographies of climate urbanism (Castán Broto and Robin 2020; Long and Rice 2019; Shi 2020) I use the notion of green structural adjustment articulated by Patrick Bigger and Sophie Webber (2020). It allows for a historically grounded but future-facing mode of analysis about how urban Africa is reimagined amid climate crisis as a site for intervention by the World Bank and the International Monetary Fund (IMF). Green structural adjustment, I will argue, represents a distinct wave of "neocolonial" control (Nkrumah 1967) vis-à-vis new forms of investment, debt, and the planning of critical infrastructure systems across urban space. I develop an anticolonial standpoint on climate urbanism. This means thinking through "a theoretical and political practice [that] illuminates an historical and analytical trajectory between the colonized world, the Third World, and the contemporary Global South" (Elam 2017).

Through reflecting on research undertaken in Mbale, Uganda, as climate change imperatives begin to shape its planning and operation of infrastructure. I consider how the waste system was restructured to capture methane emissions through a World Bank initiative of the Clean Development Mechanism (CDM), termed the Programme of Activities. Planned to test and experiment with new forms of carbon financing, technological innovation, and ways of governing both urban and atmospheric resource flows, the Uganda Municipal Waste to Compost Program might be considered an early example or even precursor of green structural adjustment and associated debt burdens in the urban response to climate change. The experience in Mbale suggests that climate urbanism is not an inherently progressive force. Low-carbon restructuring may well lead to reduced GHG emissions, but doing so has costs, often for those already marginalized and living in resource-poor contexts. Through this focus on climate urbanism in Mbale—both the everyday ways in which it is operationalized and experienced, and its financial architectures—we are forced to complicate and problematize how we understand and act on climate (in)justice (Long and Rice 2019).

Establishing what happens in Mbale is important, especially as "climate urbanism research has mainly focused on interventions led by institutional public and private actors in North American and European cities, with a few recurring examples in South America and Asia" (Robin and Castán Broto 2020, 2). While the town may be on the periphery of global knowledge production concerning climate urbanism, it is simultaneously being used as a space for "climate change experimentation" (Bulkeley and Castán Broto 2013), and I argue primarily through the rolling out of green structural adjustment by the World Bank. As Robin and Castán Broto (2020) suggest, a postcolonial approach to climate urbanism is essential to shift beyond theorizations generated from the so-called Global North to develop both situated and global conceptions of the ways climate change is predicating the restructuring of urban technoenvironments. And as I will show, a postcolonial analysis itself requires extension toward an explicitly anticolonial position.

As during the decades of Structural Adjustment Programmes (SAPs), the municipality in Mbale has little autonomy to shape its own vision of climate urbanism and is forced to follow the diktats of the World Bank and access its debt-based financial mechanisms. Mbale then offers a case in what the editors of this book suggest is the need to consider the implications of emerging governances, "where access to resilient infrastructure is mediated by global flows of capital into and out of cities" (Rice, Levenda, and Long, this volume)—and critically, whether climate urbanism as green structural adjustment has the capacity to fairly address the imperatives of climate crisis.

The chapter concludes by arguing that rather than the incurring of another generation of unmanageable debt owed by African countries, and now towns and cities to neocolonial institutions such as the World Bank, this wave of green structural adjustment should be resisted and subsequently reversed. Rather, an anticolonial understanding of the climate debt (Bond 2010; Pickering and Barry 2012)—accrued over centuries of carbon capitalism by the Global North at the cost of the Global South—becomes the primary mechanism through which a more just climate urbanism may be configured.

Climate Urbanism as Green Structural Adjustment

I have previously written about climate urbanism as generating three types of violence: the direct impacts of disasters such as floods, the secondary, mundane impacts on infrastructure leading to breakdown and disruption, and the ways responses to the crisis via low-carbon intervention cause displacements and inequalities (Silver 2018). In doing so, I highlighted the extended time/space of climate urbanism, to think through how these violent transformations to urban space connect to the (re)making of colonial, capitalist geographies (Fanon 1961; Robinson 1983). However, further work was required to extend these ideas, particularly in considering the new generation of financial mechanisms shaping the parameters of climate urbanism, and enhancing historic unequal relations between the metropole and the periphery. This line of investigation draws on the theorizations of climate urbanism that have emerged in recent years.

Long, Rice, and Levenda (2020) warn of "projects that reproduce and/or protect capitalist economies, or defensive measures done in the name of security and surveillance" (38). Ziervogel (2020) presents evidence of the problems of "neoliberal climate adaptation responses, which often entrench inequality" (734) across informal settlements. Developing new theoretical tools from the urban worlds of the South requires this critical scrutiny, but our frameworks cannot solely be focused on the contemporary urban geographies being produced. Rather, an analysis built from the experiences of these urban areas over the *longue durée* of colonial and neocolonial governance (Nkrumah 1967) becomes pressing to interrogating the logics of climate urbanism.

The recent historical starting point for the analysis presented here are the SAPs that were imposed on forty African nation-states for over three decades

(Abouharb and Cingranelli 2006) as a postcolonial form of global control (Rodney 2018). Prashad (2012) argues that "the IMF urged the indebted nations to submit themselves to complete integration in the world capitalist system" (233). Crucial to this integration was the stripping of state responsibilities and finance concerned with public welfare and environmental protection. This was based on a mantra of "trickle-down" economics, despite later empty discourse concerned with the need to "build institutions" (233).

The result of this market fetishization by the World Bank and IMF is a well-known history (Goldman 2005; Li 2007; Roy 2010). Countries implemented the stripping back of social assistance and the transfer of urban services away from state-led Keynesianism toward privatization and "public-private partnership," later described by Naomi Klein (2007) as the "shock doctrine." This included shaping the ways in which urban technoenvironments have been subsequently governed and managed. SAPs often compounded the significant development challenges the postindependence states in the so-called Global South and especially in the poorest region of sub-Saharan Africa experienced as they sought to overcome the bitter legacies of colonial exploitation, extraction, and control (Alemazung 2010).

Much like, and indeed as an extension to, the colonial, capitalist project, structural adjustment displayed clear motivations (and attendant power) to open up economies to forms of control and consequent exploitation of resources (Young and Bishop 1995; Bond 2010) crucial to the expanding frontiers of the socionatural accumulation regimes of capitalism (Moore 2015). This was achieved through means that made visible the neocolonial forms of subjugation that characterize postcolonial, center-periphery relations (Bond 2010). This control restricted the capacities of governments and people to chart alternatives to the SAPs, even when movements were elected on anti-austerity platforms. One such leader was Thomas Sankara, who contested IMF and World Bank imperialism (Sankara 1988). Sankara was clear in the ways in which the SAPs must be understood as another wave of unequal center-periphery relations in which the global North would continue to exploit Africa. He argued that "imperialism is a system of exploitation that occurs not only in the brutal form of those who come with guns to conquer territory. Imperialism often occurs in more subtle forms, a loan, food aid, blackmail. We are fighting this system that allows a handful of men on Earth to rule all of humanity."

The history of the SAPs across African towns and cities has been one of failure and social disaster (Harris and Fabricius 2005), as Sankara warned and despite debt-fueled promises of loans, aid, and assistance. There has been broad acknowledgment of this devastating period in which SAPs dominated macroeconomic planning. However, the neoliberal and neocolonial logics that guided these interventions remain stubbornly persistent and central to the ideology of the World Bank and IMF. I argue that these discredited but still important logics generate the need to question whether such a spatio-economic "fix" might be reconfigured into the governing of technoenvironments across urban space as the era of climate crisis necessitates new types of intervention.

The *longue durée* of center-periphery relations is integral to analysis of climate urbanism across urban Africa. This is because the climate crisis is enabling a new form of unequal relation vis-à-vis intervention and debt, including across the infrastructures at work in these urban spaces. Critical to this argument is the scholarship of Patrick Bigger and Sophie Webber (2020). In an intervention that can help scholars think about climate urbanism through longer histories of unequal, center-periphery relations, they propose the notion of green structural adjustment, arguing that "the World Bank is attempting to channel return-seeking investment into urban infrastructure in response to these challenges. To harness this private finance, though, cities must be reformatted in investment-friendly ways. . . . We call this rescaled and climate-inflected program of leveraged investments coupled with technical assistance Green Structural Adjustment" (2).

The work of Bigger and Webber (2020) centers the World Bank in the making of climate urbanism, as the SAPs seem increasingly repackaged around climate change imperatives. The World Bank is at the forefront of adaptation and mitigation especially in the Global South, including urban Africa, where fiscal capacity at the municipal scale is severely curtailed (in part due to decades of imposed SAPs). Green structural adjustment can be characterized as a marketized, neocolonial vision of climate urbanism through investment into infrastructure predicated on technocratic management, cost recovery, involvement of private sector actors, and potentials for new forms of debt-based relations. Bigger and Webber (2020) outline some differences between the SAPs and green structural adjustment, most notably the role of urban actors (rather than national governments) as the recipients of this flow of debt-based investment as decentralization trends continue in many African states.

The World Bank has been critical to the development of the CDM, particularly in experimentation with different types of financial intervention proceeding its "resilience turn"—as Sarah Knuth in chapter 3 of this book demonstrates. As Bigger and Webber (2020) argue, "The Bank is trying to create access to debt and investable projects where they do not currently exist, allowing urban authorities to borrow money to manage impending crises caused by rapid urbanization and climate change" (2). The CDM has become a way through which green structural adjustment has been operationalized.

I would argue it is important to connect the financial architectures that are enabling climate urbanism into the postindependence histories of African states. Doing so draws attention to how colonial control and extractions of resource hinterlands shifted to neocolonial control through the SAPs. Without this analysis, it may prove difficult to fully comprehend what climate justice requires beyond a narrow, neoliberal, and technocratic conception of new technologies and investments.

Green Structural Adjustment in Action

Mbale, Eastern Uganda, has a population expected to double to over 150,000 between 2002 and 2022 (Mbale Municipality 2010). It once had the reputation as the cleanest town in East Africa, yet presently over fifty thousand of its inhabitants are without adequate infrastructure services (UN-Habitat 2011). Reflecting broader patterns of urbanization across sub-Saharan Africa (Parnell and Pieterse 2014), Mbale's future remains in flux, shaped by a series of socionatural dynamics from floods and droughts, to the HIV epidemic and internally displaced people. Furthermore, the municipality has limited financial capacity, with no more than $5 million available in the 2015/16 budget (Ministry of Finance 2016). This restricted fiscal base can be attributed to the costs of Uganda implementing an SAP over previous decades, and the longer histories of underdevelopment in Africa (Rodney 2018).

The demands facing Mbale's municipality will be aggravated through the climate crisis as I found out when interviewing various officials in 2015 (and were subsequently categorized anonymously through various letters). Increased frequency of mudslides and growing concerns about coffee agriculture into the future (UNDP 2013) are two indicators of the climate crisis. New meteorological patterns that already include an overall increase in temperature between 1961 and 1990 show a projected rise of a further two degrees Celsius by 2060 (UNDP 2013). This temperature increase will occur alongside less frequent but more intense rainfall, and increased incidences of extreme weather (UNDP 2013). As a United Nations Environment Programme policy maker told me, climate change will have effects that will instigate "multiple vulnerabilities being faced by communities around livelihoods and particularly agriculture" (Interview G, 12.12.15). Like other urban regions, Mbale's existing infrastructure faces both the perils of underinvestment and municipal dysfunction brought on by the cumulative effects of SAPs, together with ever-growing intersections with the climate crisis.

It is in this context that we see convergence of SAPs and climate urbanism into green structural adjustment. The aim of the Uganda Municipal Waste to Compost Program initiated from 2005 was to "reduce emissions of CO_2 by recovering the organic matter from municipal solid waste as compost and avoiding methane emission" (Aenor 2010, 5). In addition to the central objective of reducing GHG emissions, underresourced municipalities such as Mbale were provided with debt-based investment for the upgrading of infrastructure to address waste management (UN-Habitat 2011). The program was planned by the World Bank, who brought in the National Environmental Management Agency (NEMA), tasked with coordinating the project across the eight municipalities. The government of Uganda financed the project through a loan (i.e., taking on debt) from the World Bank under the Environment Management and Capacity Building Project-II, providing NEMA with the $421,000 needed to establish each of the eight such facilities (Aenor 2010; UNFCC 2013). The debt for this investment from NEMA was intended to be paid back by municipalities through anticipated revenues generated through selling carbon credits on the carbon markets. In effect, it was municipalities rather than the World Bank that took on the debt and therefore the risk that came with the investment. Further costs included an upfront investment of over $85,000 from the municipality and fiscal responsibility for ongoing operation and management of the restructured waste system.

This initial investment contributed to the preparation of a suitable site, alongside the equipment required for collection and processing (primarily waste skips and a vehicle). The restructuring of the system was designed for collection to be expanded to twenty-eight points across the town for processing on a regular basis. Garbage was to be taken to a nearby facility in the industrial zone and transformed through an aerobic composting process, thereby curtailing emissions of methane. The compost was then intended to be sold to farmers, with revenue envisaged as another income stream to sustain operations and pay the debt. The system was designed with a collection capacity of 70 tons of garbage per week and to operate for up to fifteen years (Aenor 2010). It was hoped that the restructuring would provide a comprehensive solution to address the problems of waste in a low-carbon manner. As a municipal official noted, "We thought we had made a breakthrough with our waste management" (Interview E, 14.12.15). Again, as was common with the SAPs and indeed colonial governance, promises of investment and a better tomorrow from the metropole would prove to be little but empty rhetoric.

The restructuring of the waste system was conceived by the World Bank in what now looks like an attempt to experiment with a systemic shift to green structural adjustment and so-called resilience (World Bank 2010). It was established through a Programme of Activities, which is a "facility under the Clean Development Mechanism (CDM) of the Kyoto Protocol, the world's main carbon credit scheme" (Climate Focus 2011, 8). The premise of the Programme of Activities, and the key difference from earlier CDM projects, was the possibility of registering the coordinated implementation of a policy, measure, or goal that leads to GHG emission reduction across a series of sites. Once a Programme of Activities is registered, an unlimited number of component project activities can be added without undergoing the notoriously long, costly, and difficult verification process (Bryant, Dabhi, and Böhm 2015). This was mobilized by the World Bank as a strategy to "bunch" a series of GHG emission reductions across infrastructure space either within one or across regions, providing the opportunity for cities to be better integrated in the CDM and new ways to configure climate urbanism.

The pronounced motivation for the development of Programmes of Activities was to address the difficulty for urban areas in accessing investment through the carbon markets by including replicable projects with low and distributed GHG reductions into the CDM. This provides the opportunity to bunch a series of GHG emission reductions across multiple sites and support investment into low-carbon intervention unsuitable to previous verification processes (Climate Focus 2011). In Mbale the Programme of Activities was being piloted and tested, producing an important moment of what Bulkeley and Castán Broto (2013) term "climate change experimentation," specifically to find ways in which to sustain the increasingly maligned carbon markets (Bond 2016). Despite the modest GHG savings, estimated in total across the lifetime of all the municipalities, at less than 60,000 metric tons of equivalent CO_2 (Aenor 2010), it was the first Programme of Activities in urban Africa. It has therefore served its purpose as an urban-located intervention that would prefigure a larger wave of green structural adjustment through climate urbanism.

The project was registered by the CDM executive board in 2010 as part of the methodology, Avoidance of Methane Emissions through Composting (UNFCC 2013). The Programme of Activities was verified in 2013 resulting in the issuing of 16,549 certified emission reductions (or carbon credits) from the CDM (equivalent to $215,000 per annum across all eight municipalities as estimated in 2010 but turning out to be significantly less). It was the intention of the World Bank (2010) to sell these carbon credits to its own Community Development Carbon Fund. The municipality was informed during project inception of anticipated revenues of up to $26,000, but it waited for years without payment. A national government official told me, "Though we have finished the verification process we can't be sure how much and when the money will start to flow" (Interview Q, 18.12.15).

Green Structural Adjustment in the Making of Climate Urbanism

Green structural adjustment unleashes antidemocratic forces in the making of climate urbanism. In Mbale the restructuring of the waste system did not come about through an open process of planning, as one would hope. Rather, this way of managing waste vis-à-vis the CDM can be considered as imposed through a powerful transnational actor that has shaped the possibilities and scope of transformation through a low-carbon, market-oriented restructuring. This was undertaken while assigning much of the risk of failure, whether operational or financial, at the local scale. Here, we can see how the World Bank plays a pivotal role in the flows and transformation of (formerly localized and at times unstable) circulations of waste. This gave NEMA, and more so the municipality, little capacity to actively shape the design, leading to accusations of throwing their weight around by a municipal official, much like the World Bank had done for decades through the SAPs. Under the fiscal conditions faced by the local government in Uganda, the power relations in the making of climate urbanism are in this case constituted through weak municipal capacity, allowing the World Bank to impose green structural adjustment while proclaiming that the entire program is voluntary in nature.

The process and outcome are predicated on neoliberal and arguably neocolonial logics of the World Bank and its corporate and financial backers. Here we see how "climate apartheid" (Bond 2016; Dawson 2017; Rice, Long, and Levenda 2021) is not just a spatial process occurring within the city but can be understood at the global level between the powerful and the less powerful in the ways it is producing urban geographies. It will intensify the climate crisis "by normalizing the dehumanization, exploitation, and dispossession of historically oppressed communities while simultaneously promoting landscapes, infrastructures, and policies of security and isolation from climate hazards for privileged populations" (Long, Rice, and Levenda 2020, 39).

To analyze climate urbanism in Mbale means to demonstrate how green structural adjustment connects to longer histories of SAPs and takeover of urban management by the World Bank (Harris and Fabricius 2005). Furthermore, it demands new interpretations of these reconfiguring, unequal center-periphery relations. As Long (2021) argues, the "narrative is largely an export of the global North, and represents a resilience-amidst-crisis mindset that is embedded within a capitalist coloniality of power" (2).

Local policy makers were clear in their assessment of who was in control and the ways in which it has reinforced center-periphery relations between underresourced municipalities in Uganda and the World Bank. In terms of the capacity to address climate challenges and shape its own future, the municipality was sidelined. As a municipal official commented about the design of the project, "They came up with plan and we provided land, we didn't participate" (Interview J, 14.12.15). This contention was supported by an elected official in the town who questioned whether the project was the best solution for waste management and asserted that "if we were given opportunity [we] would have prioritized investment in landfill" (Interview D, 07.12.15). Instead the municipality was left with the responsibility of paying the debt and ensuring ongoing operation and maintenance. Here it is important to connect the governance of climate urbanism as constituted through green structural adjustment as World Bank–imposed economics continues to shape capacities and potentials of towns in Uganda (Branch and Mampilly 2015).

This is a fiscal context in which the government of Uganda (2010) has "adopted a policy of public-private partnership as a tool for the provision of public services and public infrastructure" (3). This turn toward the private sector across municipalities has been intensified through ongoing World Bank–led reforms, especially concerning procurement from the late 1990s onward that have effectively privatized many municipal services (Guma 2016). In this case study, the public-private partnership, a standard outcome of neoliberal governance is extended into responses to the climate crisis. This is undertaken through mobilizing low-carbon concerns, alongside the promise of (much needed) investment and future revenues to local partners. The logics of climate urbanism through green structural adjustment meant that the municipality was unable to operate the system as a public service. It was forced to use a private operator and unable to participate in decision-making about the conditions of investment. As a senior official contended, "the contractors and the management of this contract is one of many challenges; he takes the money but when you look at what the outputs are [they are] just not there" (Interview F, 12.12.15).

There was criticism beyond the municipality concerning governance and the power relations it served to reinforce. The anger was articulated by the National Slum Dwellers Federation of Uganda as they claimed to have been ignored during planning and implementation; this was coupled with the failure to deliver promised community benefits. As far as the federation was concerned, these benefits were broken promises with little being delivered. Furthermore, failure to operate the system as intended meant that underserviced neighborhoods continued to experience waste management issues, resulting in vulnerabilities to ever-present hazards such as cholera. The waste-to-compost project was supposedly designed to address local imperatives, but it became evident for social movements and civic organizations that these were simply not being addressed. The implications of this are serious, in that waste would continue to create precarious socio-ecological conditions for some of Mbale's poorest residents. As Ziervogel (2020) argues, "The sad fact is that the urban poor have often been the losers in urban adaptation responses, and will continue to be so in the future" (734). Mbale shows this perspective to be equally true for urban mitigation responses. Other potentially democratic or at least "participatory" governance possibilities for Mbale to improve waste management while addressing the climate crisis were foreclosed through the imposition of this green structural adjustment. A municipal official explained, "This is a World Bank project with its own set of guidelines, they have their own ways" (Interview J, 14.12.15).

The fiscal conditions and socio-ecological challenges faced by the municipality, alongside growing imperatives to address the climate crisis, became productive for the World Bank by allowing the institution to impose its own vision of climate urbanism. Here Mbale provides an example of what Rice (2014) terms the "territorial politics of carbon" (385) as it becomes entwined in longer histories of SAPs and neocolonial control across East Africa. The process and outcome have been both predicated on and helped expand neoliberal logics of infrastructure operation (Graham and Marvin 2001) into the era of climate crisis (Long and Rice 2019; North, Nurse, and Barker 2017). It has meant that a neocolonial actor became central to the shaping of climate urbanism and to the management of circulations of urban resources including waste and GHG emissions. Simultaneously, it left the municipality with the difficulties of paying the debt and maintaining a new investment while dealing with procurement guidance and legislation that restricted its ability to operate this new system on its own terms. Such imposition of a particular set of (green) neoliberal logics, as part of what Obeng-Odoom (2014) terms a "broader process of global change towards marketizing the environment" (129), raises important questions about the capacity of local authorities to govern climate urbanism. Here a democratic process to prioritize which investments would be best suited to waste management and curtailing GHG emissions was nonexistent. This led to the taking on of a debt and particular operational logics, in this case marketized and low-carbon, orienting the nature of this restructuring while allowing for little local scrutiny or decision-making (Bigger and Webber 2020).

Against Green Structural Adjustment

Unable to rely on the fiscal capacity of better-resourced cities, and battered by years of underinvestment from the SAPs, municipalities such as Mbale remain at once hesitant and overawed to impositions by governance actors already deeply implicated in decades of control that have perpetuated uneven center-periphery relations (Briggs and Yeboah 2001). This way of governing infrastructure as climate urbanism has in turn led to the commodification of the waste and GHG "commons" of the town and new forms of financialization (in the shape of the Programme of Activities). I have demonstrated how green structural adjustment serves to usher in a new era of neoliberal urban resource management (Heynen, Kaika, and Swyngedouw 2006), or as the editors of this book argue, simply establishes "new mechanisms that ultimately entrench old injustices."

Climate urbanism has created reconfigured, unequal power relations with implications for various local actors. For waste pickers and workers, this includes struggles to survive as waste commons and wages are stolen. For surrounding communities, this translates to ongoing issues in the management of waste, and for the municipality it means an inability to address multiple, unstable socio-ecological futures in an era of climate crisis. Evidence from Mbale illustrates how new forms of "accumulation by dispossession" (Harvey 2005) and enclosure might become materialized through a climate urbanism that itself is structured through the logics of green structural adjustment. Simultaneously, the livelihood strategies of the poor (in this case, waste pickers) are deemed as blocking progress in addressing the crisis. Climate urbanism is in part about turning GHG emissions and circulations of waste in urban areas into exchange value that travel beyond municipal boundaries. Potential displacement draws attention to how poor people are exposed to new precarity through climate urbanism. As I have set out and as Sarah Knuth in the previous chapter convincingly demonstrates, contemporary analysis of the financing of new climate interventions requires thinking across the longer histories of colonial capitalism and appropriation of people and uncapitalized natures (Moore 2015).

As Whitington (2012) has cautioned, the creation of these new (carbon) markets is predicated on the wholesale transfer (or theft) of emission rights from countries such as Uganda to the Global North. This carbon dispossession (Böhm, Misoczky, and Moog 2012) as a form of climate urbanism displays similarities to the colonial project that plundered the resources of countries such as Uganda, calling attention to the continued inequalities of center-periphery relations, (neo)coloniality, and socio-environmental violence. The experience in Mbale shows that the violence of climate change will not only unfold through multiple and unequal socio-environmental disasters brought forth through GHG emission–generated changes to the atmosphere (Parenti 2011). The global response to the crisis, increasingly centered on climate urbanism, seems likely to reinforce unjust relations between center and periphery. This is likely to occur due to both the ineffectiveness of actions to significantly reduce GHG emissions and the inequalities and power relations generated through green structural adjustment and the debts that result.

Paying attention to out-of-the-way places like Mbale is important as climate urbanism proceeds across and through infrastructure. What happens on the "periphery" may prefigure the climate urbanism futures (both within and exceeding geographical containers such as the Global South). The elsewheres of global urbanization become spaces for experimenting with financing initiatives such as the CDM, within the broader logics of green structural adjustment. Climate urbanism should not just be concerned with the ways in which urban space is reshaped. Rather, it has centered new forms of neoliberal "green" logic and circulations of carbon capital becoming materialized as a wave of green structural adjustment across the urban scale. Global institutions exert power and commodify and financialize local resources to pilot, test, and develop neoliberal and neocolonial operation of infrastructure as a response to climate crisis. This is the financial architecture of climate urbanism, in which attempts to better manage GHG emissions entwine urban areas with the logics of market-based climate action, generating new forms of inequality and power relations, primarily through debt.

Let's return to Thomas Sankara 1987 to consider the debt-based nature of Green Structural Adjustment, and its financial models such as the Clean Development Mechanism. He argued at the Organization of African Unity, Addis Ababa in 1987: "We think that debt has to be seen from the standpoint of its origins. Debt's origins come from colonialism's origins. Those who lend us money are those who had colonized us before. Under its current form, that is imperialism-controlled, debt is a cleverly managed re-conquest of Africa, aiming at subjugating its growth and development through foreign rules. Thus, each one of us becomes the financial slave, which is to say a true slave." Sankara's anticolonial argument poses the question in how we think about these climate-changed futures, and the likely dominance of green structural adjustment in shaping climate urbanism—specifically, who has debt to whom? Such an exercise opens up the notion that Global North countries and institutions such as the World Bank must pay a "climate debt" (Bond 2010; Pickering and Barry 2012) predicated on a reparative way to finance climate urbanism across sub-Saharan urban Africa, and as an explicit break from centuries of socionatural exploitation and extraction. This is financing anchored in wider understandings concerning ecological debt (Goeminne and Paredis 2010) and of payment for the historical, colonial, and capitalist pollution of the atmosphere. Such a reparative model offers an alternative to the failed logics of markets, and could upend unequal, center-periphery inequalities that continue to characterize relations through the implementation of green structural adjustment.

Building on Sankara's understanding, climate justice movements must up the pressure on governments in the Global North to make payment for the climate debt through reparative mechanisms such as the Greenhouse Development Rights framework or financial transaction taxes (Darvas and Weizsäcker 2011). Developing alternative forms of distribution of global financial resources vis-à-vis reparations for intensifying climate crisis, as well as acknowledging the climate debt, would not automatically reconfigure a more just climate urbanism. However, as Max Ajl (2021) argues in relation to the "People's Agreement of Cochabamba," such decolonial praxis, built on acknowledgment of climate debts, might open up "planks of a Southern platform for ecological revolution" (11). Linking these reparative circulations of finance into Green New Deal–style transformations of towns and cities (Ajl 2021; Pettifor 2020) would generate spaces for the emerging alliances of southern-based environmental and social movements, marginalized and vulnerable groups, and the growing numbers of climate-savvy young people. Doing so would certainly echo the mobilizations that Sankara and his revolutionary allies enabled during the Burkina revolution, including through environmental action (Murrey 2018). New finance would have to flow not only to municipal budgets but also to initiatives such as basic income programs (Van Parijs 2017; Wright 2004) that focus on emerging community-level responses, which may help support and empower resource-poor neighborhoods to self-organize for the climate crisis. Here we all have a role to play in supporting attempts to develop a more just climate urbanism from below, a collective struggle to configure a future that acknowledges and attempts to repair the damage caused by longer histories of slavery, underdevelopment, and resource extractions and shift away from a new wave of neocolonial control imposed through green structural adjustment.

REFERENCES

Abouharb, M. R., and Cingranelli, D. L. (2006). The human rights effects of World Bank structural adjustment, 1981–2000. *International Studies Quarterly*, 50(2), 233–262.

Aenor. (2010). Validation Report of the POA for Uganda Waste to Compost Program. http://tinyurl.com/h9hnawr.

Ajl, M. (2021). *A People's Green New Deal*. Pluto Press.

Alemazung, J. A. (2010). Post-colonial colonialism: An analysis of international factors and actors marring African socio-economic and political development. *Journal of Pan African Studies*, 3(10), 62–84.

Bigger, P., and Webber, S. (2020). Green structural adjustment in the World Bank's resilient city. *Annals of the American Association of Geographers*, 111(1), 36–51.

Böhm, S., Misoczky, M. C., and Moog, S. (2012). Greening capitalism? A Marxist critique of carbon markets. *Organization Studies*, 33(11), 1617–1638.

Bond, P. (2010). Climate debt owed to Africa: What to demand and how to collect?. *African Journal of Science, Technology, Innovation and Development*, 2(1), 83–113.

———. (2016). Who wins from climate apartheid?: African climate justice narratives about the Paris COP21. *New Politics*, 15(4), 83.

Branch, A., and Mampilly, Z. (2015). *Africa Uprising: Popular Protest and Political Change*. Zed Books.

Briggs, J., and Yeboah, I. E. A. (2001). Structural adjustment and the contemporary sub-Saharan African city. *Area*, 33(1), 18–26.

Bryant, G., Dabhi, S., and Böhm, S. (2015). "Fixing" the climate crisis: Capital, states, and carbon offsetting in India. *Environment and Planning A: Nature and Space*, 47(10), 2047–2063.

Bulkeley, H., and Castán Broto, V. (2013). Government by experiment? Global cities and the governing of climate change. *Transactions of the Institute of British Geographers*, 38(3), 361–375.

Castán Broto, V., and Robin, E. (2020). Climate urbanism as critical urban theory. *Urban Geography*, 42(6), 715–720.

Castán Broto, V., Robin, E., and While, A. (Eds.). (2020). *Climate Urbanism: Towards a Critical Research Agenda*. Palgrave Macmillan.

Climate Focus. (2011). The Handbook for Programs of Activities. www.climatefocus.com/sites/default/files/handbook_for_programme_of_activities_2nd_edition.pdf.

Darvas, Z., and Weizsäcker, J. (2011). Financial transaction tax: Small is beautiful. *Society and Economy*, 33(3), 449–473.

Dawson, A. (2017). *Extreme Cities: The Peril and Promise of Urban Life in the Age of Climate Change*. Verso.

Elam, J. D. (2017). "Anticolonialism." Global South Studies: A Collective Publication with The Global South. https://globalsouthstudies.as.virginia.edu/key-concepts/anticolonialism.

Fanon, F. (1961). *The Wretched of the Earth*, tr. Constance Farrington, preface by Jean-Paul Sartre. Grove Weidenfeld.

Goeminne, G., and Paredis, E. (2010). The concept of ecological debt: Some steps towards an enriched sustainability paradigm. *Environment, Development and Sustainability*, 12(5), 691–712.

Goldman, M. (2005). *Imperial Nature: The World Bank and Struggles for Social Justice in the Age of Globalization*. Yale University Press.

Government of Uganda. (2010). Public-Private Partnership Framework Policy. Department of Finance, Planning and Economic Development, Kampala.

Graham, S., and Marvin, S. (2001). *Splintering Urbanism: Networked Infrastructures, Technological Mobilities and the Urban Condition*. Psychology Press.

Guma, P. K. (2016). The governance and politics of urban space in the postcolonial city: Kampala, Nairobi and Dar es Salaam. *Africa Review*, 8(1), 31–43.

Harris, N., and Fabricius, I. (Eds.). (2005). *Cities and Structural Adjustment*. Routledge.

Harvey, D. (2005). *The New Imperialism*. Oxford University Press.

Heynen, N., Kaika, M., and Swyngedouw, E. (Eds.). (2006). *In the Nature of Cities: Urban Political Ecology and the Politics of Urban Metabolism*. Taylor & Francis.

Klein, N. (2007). *The Shock Doctrine: The Rise of Disaster Capitalism*. Macmillan.

Li, T. M. (2007). *The Will to Improve: Governmentality, Development, and the Practice of Politics*. Duke University Press.

Long, J. (2021). Crisis capitalism and climate finance: The framing, monetizing, and orchestration of resilience-amidst-crisis. *Politics and Governance*, 9(2), 51–63.

Long, J., and Rice, J. L. (2019). From sustainable urbanism to climate urbanism. *Urban Studies*, 56(5), 992–1008.

Long, J., Rice, J. L., and Levenda, A. (2020). Climate urbanism and the implications for climate apartheid. In V. Castán Broto, E. Robin, and A. While (Eds.), *Climate Urbanism: Towards a Critical Research Agenda* (pp. 31–49). Palgrave Macmillan.

Mbale Municipality. (2010). Mbale Development Plan.

Ministry of Finance. (2016). Local Government Performance Contract FY 2015/16 for Mbale. http://budget.go.ug/budget/sites/default/files/Indivisual%20LG%20Budgets/Mbale%20MC%20Final%20Form%20B.pdf.

Moore, J. (2015). Capitalism in the Web of Life: Ecology and the Accumulation of Capital. Verso Books.

Murrey, A. (Ed.). (2018). *A Certain Amount of Madness: The Life, Politics and Legacies of Thomas Sankara*. Pluto Books.

Nkrumah, K. (1967). *Neo-colonialism: The Last Stage of Imperialism*. PANF.

North, P., Nurse, A., and Barker, T. (2017). The neoliberalization of climate? Progressing climate policy under austerity urbanism. *Environment and Planning A: Economy and Space*, 49(8), 1797–1815.

Obeng-Odoom, F. (2014). Green neoliberalism: Recycling and sustainable urban development in Sekondi-Takoradi. *Habitat International*, 41, 129–134.

Parenti, C. (2011). *Tropic of Chaos: Climate Change and the New Geography of Violence*. Bold Type Books.

Parnell, S., and Pieterse, E. (2014). *Africa's Urban Revolution*. Zed Books.

Pettifor, Ann. (2020). *The Case for the Green New Deal*. Verso.

Pickering, J., and Barry, C. (2012). On the concept of climate debt: Its moral and political value. *Critical Review of International Social and Political Philosophy*, 15(5), 667–685.

Prashad, V. (2012). *The Poorer Nations: A Possible History of the Global South*. Verso.

Rice, J. L. (2014). An urban political ecology of climate change governance. *Geography Compass*, 8(6), 381–394.

Rice, J. L., Long, J., and Levenda, A. (2021). Against climate apartheid: Confronting the persistent legacies of expendability for climate justice. *Environment and Planning E: Nature and Space*. https://doi.org/10.1177/2514848621999286.

Robinson, C. J. (1983). *Black Marxism: The Making of the Black Radical Tradition*. University of North Carolina Press.

Rodney, W. (2018). *How Europe Underdeveloped Africa*. Verso.

Roy, A. (2010). *Poverty Capital: Microfinance and the Making of Development*. Routledge.

Sankara, T. (1988). *Thomas Sankara Speaks: The Burkina Faso Revolution, 1983–87*. Pathfinder.

Shi, L. (2020). The new climate urbanism: Old capitalism with climate characteristics. In V. Castán Broto, E. Robin, and A. While (Eds.), *Climate Urbanism: Towards a Critical Research Agenda* (pp. 51–65). Palgrave Macmillan.

Silver, J. (2018). Suffocating cities: Urban political ecology and climate change as social-ecological violence. In H. Ernstson and E. Swyngedouw (Eds.), *Urban Political Ecology in the Anthropo-obscene: Interruptions and Possibilities* (pp. 129–146). Routledge.

UNFCC. (2013). Monitoring report form for CDM program of activities. https://cdm.unfccc.int/ProgrammeOfActivities/poa_db/JL4B8R2DKF9oNE6YXCVOQ3MWSGT5UA/view.

UNDP Uganda. (2013). Climate Profiles and Climate Change Vulnerability of the Mbale Region of Uganda: Policy Brief. UNDP Uganda.

UN-Habitat. (2011). *Cities and Climate Change: Global Report on Human Settlements*. Earthscan.

Van Parijs, P. (2017). *Basic Income: A Radical Proposal for a Free Society and a Sane Economy*. Harvard University Press.

Whitington, J. (2012). The prey of uncertainty: Climate change as opportunity. *Ephemera: Theory and Politics in Organization*, 12, 113–137.

World Bank. (2010). *A City-wide Approach to Carbon Finance*. Carbon Partnership Facility Innovations Series. World Bank.

Wright, E. O. (2004). Basic income, stakeholder grants, and class analysis. *Politics and Society*, 32(1), 79–87.

Young, C. E. F., and Bishop, J. (1995). Adjustment Policies and the Environment: A Critical Review of the Literature. International Institute for Environment and Development, London.

Ziervogel, G. (2020). Climate urbanism through the lens of informal settlements. *Urban Geography*, 42(6), 733–737.

Climate Praxis for the Just City

CHAPTER 5

Leveraging Urban Climate Action for Transformative Social Justice

JOAN FITZGERALD, GLORIA SCHMITZ,
AND JENNIE C. STEPHENS

This chapter highlights the opportunities and challenges of urban climate action that reduces racial and economic disparities. The climate crisis and the COVID-19 pandemic reveal injustices and create an urgency to advance social justice and equity in climate action. To advance the climate-just-cities agenda, we frame this chapter around two interrelated questions: How do we assess the extent to which social justice, racial justice, and economic justice are key principles underpinning a city's climate action agenda, and what barriers do cities face when advancing social justice in implementation of their climate plans? Examining these questions is a first step in revealing the potential of urban climate action to reverse growing urban inequities and prevent urban climate action from perpetuating vulnerabilities and injustices. This inquiry introduces a methodology for scholars and practitioners to assess how equity is incorporated into urban climate practice.

We begin with a review of urban climate justice movements and discuss how equity and social justice have been measured in urban climate action. Several researchers have conducted content analyses of climate action plans to reveal the extent that equity has been incorporated. We argue that focusing exclusively on a city's climate action plans is limiting because other related city plans may integrate climate action with equity and social justice. To examine this premise, we review Boston's and Seattle's climate action and related plans to evaluate the extent to which they have integrated social justice. We then explore barriers that cities face linking social justice with climate action. We conclude by reflecting on how future urban climate planning might more effectively leverage the transformative potential of advancing social justice, and how advocates of social, racial, and economic justice might engage with climate action planning processes in cities.

The Disconnect between the Urban Climate Justice
Movement and Urban Climate Action

The idea of a just urban transition is gaining traction internationally (Hughes and Hoffman 2020), and urban climate justice is a key component of that. The climate-just cities framework calls for integrating considerations of social, environmental, and ecological justice into urban climate action planning (Steele et al. 2012). As noted in the introduction to this book, it calls for urban climate security and equality in protecting all people from climate hazards while also including all residents in the benefits of mitigation.

This movement follows an urban environmental justice (EJ) movement of more than two decades in the United States. EJ Organizations in many cities have documented the historical exposure of communities of color and low-income neighborhoods to environmental harms such as highways, incinerators, wastewater treatment plants, landfills, and other undesirable land uses (see Tessum et al. 2021; Mohai and Saha 2015; Bullard 1994). These organizations have had varying degrees of success in convincing city government to redress these harms (Hess and Satcher 2019; Agyeman et al. 2016).

Despite these "bottom-up" calls for climate and environmental justice, city climate action planning has not, for the most part, incorporated equity/justice into the agenda. Most climate mitigation plans focus on reducing emissions without regard to the needs of frontline communities. Urban renewable energy policies and energy efficiency programs, for example, often underserve or exclude low-income, underserved, and marginalized households. There are both racial/ethnic and socioeconomic disparities in access to energy-efficient residential technology (Lukanov and Krieger 2019; Reames, Reiner, and Stacey 2018; Carley and Konisky 2020). Likewise, programs to expand renewable energy adoption cater to homeowners who have the capacity to pay the upfront costs needed to take advantage of subsidies (Fitzgerald 2020; Stephens 2020). Workforce training and access to renewable energy and energy efficiency jobs are also often inaccessible to many, perpetuating a lack of diversity in the clean energy workforce (Muro et al. 2019).

A growing literature critiques most urban adaptation planning for its lack of attention to issues of climate equity, social justice, and systemic transformations (Anguelovski, Connolly, and Brand 2018; Fainstein 2018; Harris, Chu, and Ziervogel 2017; Kaika 2017; Meerow, Pajouhesh, and Miller 2019; Ziervogel et al. 2017). Evidence is accumulating that adaptation and resilience planning in many cities focuses on wealthy neighborhoods, often ignoring and/or harming low-income neighborhoods and communities of color (Fitzgibbons and Mitchell 2019; Béné et al. 2018; Meerow and Newell 2017; Landry, Dupras, and Messier 2020).

Cities increasingly are being pressured to address environmental justice in climate action. In some cities, attention to climate justice is coming from staff of offices of sustainability (Fitzgerald 2022). As these efforts evolve, it is important that we evaluate their effectiveness. To this end, we offer an initial approach to measuring the extent to which equity is meaningfully incorporated into urban climate action plans.

Measuring Equity in Urban Climate Action

Previous research assessing equity in urban climate planning has revealed that environmental justice is not a focus (Finn and McCormick 2011). Bulkeley and Castán Broto's (2013) international survey of more than a hundred cities implementing climate change experiments found that only about one-fourth included environmental justice concerns. Schrock, Bassett, and Green (2015) and Warner (2010) concluded that equity was a relatively low priority of the twenty-eight U.S. cities they examined. Following Bullard (1994), equity was defined in three ways: procedural (public participation), geographic (across neighborhoods), and social (race, ethnicity, income) (Schrock, Bassett, and Green 2015, 286).

Building on Schrock, Bassett, and Green (2015), Waud (2017) examined whether social justice is identified as a problem, goal, or action in the climate action plans of nineteen Carbon Neutral Cities Alliance member cities. Waud differentiates three categories of inclusion: absent, isolated, or integrated. The isolated category refers to plans in which equity considerations are contained in a separate section, with little or no integration with action items, while integrated plans have equity as a key principle integrated into all components of the plan and its action steps. Waud's analysis also examined whether plan sections on equity included multiple themes such as social development, economic inclusion, and health equity. Those plans mentioning more than one theme were categorized as integrated. Waud found that six of the nine U.S. city plans she reviewed had a fully integrated equity agenda.

Two studies have examined how progress toward equity goals is measured in urban climate action plans (York and Jarrah 2020). Of sixty-six cities worldwide, only two, Chicago and New York, established a measure for progress toward energy equity, defined by reducing the energy burden of low-income groups and improving energy affordability, and only eight of the cities actively measure equity as it relates to energy and transportation implementation (York and Jarrah 2020, 23). Fitzgerald (2022) found that five cities that recently updated their climate plans to focus on equity had adequate metrics and strategies for ensuring that equity goals would be met during implementation.

A limitation of all these studies is that they only assess the city's climate action plans, assuming that they are unrelated to other plans. Given that most cities have multiple plans that are integrated with their climate action plans, the lack of an integrated assessment of all of a city's planning documents has led to an incomplete understanding of how cities are incorporating racial and social equity goals into their planning. Recognizing this gap, we turn to evaluating how social justice, racial justice, and economic justice are integrated into the climate action and related plans of two large cities recognized for innovation in climate action: Boston, Massachusetts, and Seattle, Washington.

Evaluating the Integration of Social Justice in Urban Climate Planning in Boston and Seattle

We employed a comparative case study of Boston and Seattle to develop a framework for measuring the extent to which social justice and equity goals are established in multiple climate-related plans and the extent to which measurement of progress is ongoing. Boston is an ideal city for this analysis because it is considered a leading city on climate action and has high levels of income inequality and neighborhood segregation based on income and race, which it is trying to ameliorate with various policies. Seattle is also a leading city on climate action and is recognized for its equity and climate justice initiatives.

Boston was rated first in the nation for the past five years—with the state of Massachusetts ranking first since 2011—on the American Council for an Energy Efficient Economy's Clean Energy and Energy Efficiency Scorecards (Samarripas et al. 2021). This indicator measures energy efficiency policy and programs and provides recommendations for improving energy policy. Another Boston accolade is being ranked an "A-list" city by the Climate Disclosure Project (CDP), a nonprofit environmental disclosure system for cities, states, and regions, for its ambitious and comprehensive climate action planning and greenhouse gas emission legislation. As a member of the Carbon Neutral Cities Alliance (CNCA) since its formation in 2014, Boston has committed to equity as a guiding principle in its carbon reduction strategies. What remains to be examined is the extent to which Boston has delivered on this commitment.

Boston was ranked the worst city in the nation on income inequality by the

Brookings Institution in 2014, with top income earners making eighteen times as much as those in the bottom. The gap has narrowed since then, dropping to the seventh most unequal city in 2016, a reflection of worsening conditions in other cities more than improvements in Boston. The 95th percentile make ten times as much as the 20th percentile (Berube 2018). This income inequality is reflected in neighborhoods, with wide income inequality among neighborhoods in Boston and among most suburbs of the metropolitan area (Esri 2017). While climate action planning cannot be expected to ameliorate such gaping inequality, a political commitment to doing so should be reflected in the climate action and related plans.

Seattle also is well respected for its climate mitigation planning initiatives. The ACEEE rated Seattle third in the United States for energy efficiency in 2020 (Ribeiro et al. 2020). In addition, Seattle was named an A-list city for 2019 by CDP for its ambitious target of 100 percent emissions reductions by 2050.[1] Seattle is also a member of the Carbon Neutral Cities Alliance and has reduced its overall emissions by 6 percent since 2008 (CNCA 2020).

On the inequality front, Seattle was ranked the eighteenth worst city for income inequality by the Brookings Institute in 2017, with its top income earners ranking the third richest in the United States. This income inequality is exacerbated by low average wages, with Seattle ranking sixth out of nine West Coast cities (Balk 2017). Although urban planning cannot close Seattle's wage gap, these statistics demonstrate the importance of creating climate change plans that include detailed climate justice initiatives for its socially vulnerable residents, which Seattle has been actively trying to incorporate.

To evaluate the plans, we conducted a content analysis. Following Guyadeen and Seasons (2016, 100), this analysis assesses the equity and distributional equity goals of a plan and whether it identifies methodology for evaluating inclusion and equity goals during implementation. This plan and planning evaluation is distinct from formative and summative evaluations of implementation (see Patton 2002). Seasons and a research assistant independently conducted a content analysis of the climate action plans of both cities. The authors used the Adobe "search PDF" function to search the plans for key terms and phrases related to equity in each section of the plans. These terms were environmental justice, environmental equity, low income, racial equity, socially vulnerable populations, disparity, race/racial, ethnicity, green gentrification, and social resilience. As the search identified each use of the term, we analyzed the context and counted it if it was used to justify a goal or as part of a goal. Next we determined whether the plans identified outcome measures for equity goals and a public reporting system. Based on this analysis, the plans were categorized as "Absent," "Isolated," "Integrated," and "Evaluated." We categorized the reports to align with Waud's (2018) definition of equity as "Isolated" if equity concerns were not present in most of the report areas. "Integrated" plans incorporated equity into multiple report areas but failed to show any evaluation metrics or progress tracking of equity goals. We added the category of "Evaluated" if plans mentioned equity in all sections and showed progress toward employing equity tools or provided evaluation metrics, such as equity and racial justice toolkits.

We then completed the same analysis for other city or county plans complementary to or integrated with the climate action plans. For Boston, we identified nine related plans that are also complementary to Boston's recent comprehensive plan, Imagine Boston 2030 (Boston Planning and Development Agency 2017). For Seattle, we identified five related plans from both the City of Seattle and King County that are related to its climate action plan (City of Seattle 2018). We searched for mentions of justice-related keywords and phrases in each of these plans. In addition, we examined the extent to which these plans connected to the climate action plan.

BOSTON'S AND SEATTLE'S CLIMATE ACTION PLANS

We characterized both the Boston 2014 climate action plan and Seattle 2013 climate action plan as "Integrated" because entire sections discussed equity, equity was mentioned in most of the sections, and/or explicit goals about equitable distribution of and access to resources were included. The Boston Climate Resilience plan's initiatives state explicit goals about equitable distribution of resources, including, "The City should work to establish a network of neighborhood-level volunteers who check on socially vulnerable populations" (11) and "The City should prioritize developing retrofit financing pathways for buildings that serve vulnerable populations" (35).

The plans fell short, however, of identifying specific equity goals with milestones for achieving them and measuring progress over time. For example, one of Seattle's equity goals was to increase access to energy-efficient residential units, but there is no metric for measurement, nor a time line for building a defined number of units. In Boston, similar equity goals were mentioned without specific time lines and evaluation metrics for implementation. For example, Boston's Carbon Free Social Equity Report (Boston Green Ribbon Commission 2019) mentions the need for deep energy retrofit requirements and aims at "creating financial mechanisms to reduce cost burden" (32). However, there is no proposed time line for this, nor more specific information given on how this goal will be accomplished.

Boston's 2014 climate action plan defines social equity as a cross-cutting theme. It states that the impacts of climate change should not be disproportionally borne by low-income and minority communities, and that these communities share in the benefits of the city's climate mitigation and adaptation strategies. For example, the Boston Housing Authority's Strategic Sustainability Plan (2014), states that it is "committed to soliciting and incorporating the diverse perspectives of its constituents in all aspects of its operations . . . and seeks to involve its residents and staff in order to identify the best strategies . . . and to facilitate successful plan implementation" (5). Another example mentioned in Greenovate Boston's Climate Action Plan update (2014) is that eleven thousand weatherizations or heating systems upgrades had been undertaken in low-income households.

The Seattle Climate Action Plan (Foster et al. 2013) contains references to equity in its sections on transportation, infrastructure, health and safety, waste, and resilience. This plan contains multiple references to the equity-related terms in many of its sections, but it did not include an evaluation metric or specific time line for the implementation of its equity goals. For example, in its section on transportation it mentions "Transportation and Planning & Development, which includes health, safety and equity outcomes in transportation." However, this section is not specific on how it will include equity in its transportation policy. The report does acknowledge the disproportionate impacts of climate change on socially vulnerable populations in its 2030 vision.

BOSTON'S ADDITIONAL CLIMATE-RELATED PLANS

Boston has nine other climate-related reports or plans published since the 2014 climate action plan (table 5.1), each with multiple dimensions of equity included. Most of the plans fell into the "Integrated" category.

The Strategic Sustainability Plan (Boston Housing Authority 2014) was placed in the "Isolated" category because it contained very few mentions of equity and racial justice, while the Housing a Changing City Boston 2030 (City of Boston 2019b) plan was "Evaluated" because it discussed the implementation and evaluation of equity goals in nearly all of its sections. Go Boston 2030 (Boston Transportation Department 2017) has specific equity goals in its public transit expansion plan. For example, Boston aims to create separated bus rapid transit lanes in low-income areas to decrease wait times and increase on-time bus arrivals. Carbon Free Boston's Social Equity Report (Boston Green Ribbon Commission 2019) includes the goal of placing rooftop solar panels on residential buildings in socially vulnerable communities, but there are no corresponding measures to reporting progress. This was also true of the five other

TABLE 5.1. Inclusion of equity in Boston's climate reports

Report	Absent	Isolated	Integrated	Evaluated
Boston Climate Action Plan Update (2014)			X	
Go Boston 2030 (2017)			X	
Imagine Boston 2030 (2017)			X	
Carbon Free Boston: Social Equity Report (2019)			X	
Climate Resilience Initiatives (2019)			X	
Mayor's Food Access Agenda (2019)			X	
Climate Vulnerability Assessment (2019)			X	
Economic Inclusion + Equity Agenda (2017)			X	
Strategic Sustainability Plan (2014)		X		
Housing a Changing City Boston 2030: 2018 Update, Q1 Progress Report (2019)				X

TABLE 5.2. Equity strategies in Boston's planning reports

Report Focus	Strategies
Affordable Housing	▪ Provide more green, affordable housing (e.g., low-income Passive House construction) ▪ Promote green leases ▪ Increase green building standards in housing operated by the Boston Housing Authority ▪ Target energy retrofits to low-income housing
Neighborhood Resilience	▪ Implement heat island strategy, including green stormwater management, tree planting, and more green space in low-income neighborhoods ▪ Increase protection from sea level rise ▪ Provide greater access to healthy food
Renewable Energy	▪ Promote community solar projects ▪ Identify low-income neighborhoods for microgrid development
Transportation	▪ Increase accessibility to transit in underserved neighborhoods ▪ Develop subsidized fare programs ▪ Develop a subsidized EV sharing program to serve low-income neighborhoods ▪ Expand "Complete Streets" implementation in low-income neighborhoods
Green Jobs	▪ Increase green job creation and training efforts (e.g., energy efficiency retrofitting and related construction, solar installation, green infrastructure, renewable energy, and green technology manufacturing supply chain) ▪ Provide technical assistance to firms to transition into making greener products with cleaner methods

equity goals identified in the plan. Most of the reports integrate equity goals (table 5.2). The equity goals, however, seldom identify specific or quantifiable actions. The Housing a Changing City report, for example, says the city will "protect vulnerable residents and communities by preparing Boston's housing stock for climate change including sea-level rise, extreme heat and precipitation, and natural disasters" (30). Still, the city is on record for equity goals and for reporting on achieving them. A goal of Go Boston 2030 is that "every Bostonian will live within a 10-minute walk of transit, bikeshare, and carshare," but there is no stated time line on when this goal will be accomplished.

These related plans complement the social vulnerability indicators in the Climate Ready Boston plan that include access to public transportation, urban greenspace, affordable housing, and food. The plans integrate equity goals but do not identify specific frontline communities that should be prioritized. We conclude that Boston needs to initiate a process for identifying a time line for achieving specific goals and the communities that should be prioritized. This process has to have a higher level of participation from members of frontline communities.

SEATTLE'S ADDITIONAL CLIMATE-RELATED PLANS

Seattle has half as many climate-related plans as Boston. We placed two of its plans in the "Evaluated" category and one plan in the "Absent" category (table 5.3).

We located the 2018 climate action plan in the "Absent" category because it does not contain specific references to equity, instead focusing on the technical aspects of Seattle's GHG emissions and the accomplishments in different sectors, including transportation, carbon pricing, and building retrofits. Its lack of equity references is explained in the city's decision to create a freestanding report on equity, the Equity and Environment Agenda (Seattle Office of Sustainability and Environment 2017). This report makes specific references to how equity has been included in its goals and accomplishments. For example, the plan will use the existing equity assessments to create a cumulative impacts assessment methodology that examines how racial discrimination, lack of economic opportunities, and other social inequities negatively impact socially vulnerable populations (27). This assessment methodology will identify key environmental justice issues to address and track the progress of equity issues.

In addition, the King County Strategic Climate Action Plan (2020) includes equity goals throughout (table 5.4). We placed this plan in the "Evaluated" category because it includes performance measures for evaluating progress (7).

TABLE 5.3. Inclusion of equity in Seattle's climate reports

Report	Absent	Isolated	Integrated	Evaluated
Seattle Climate Action Plan (2013)			X	
Seattle City Light Transportation Electrification Strategy (2019)			X	
Seattle Climate Action (2018)	X			
Equity & Environment Agenda (2017)				X
King County Strategic Climate Action Plan (2020)				X

TABLE 5.4. Equity strategies by report areas for Seattle

Report Focus	Strategies
Affordable Housing	▪ Increase construction of affordable housing ▪ Prioritize affordability of homes within half a mile of frequent transit service ▪ Increase tenant protections ▪ Protect communities of color and low-income communities from displacement by gentrification
Neighborhood Resilience	▪ Enhance equity in climate change preparedness ▪ Conduct assessments of climate change impacts on socially vulnerable communities
Renewable Energy	▪ Reduce greenhouse gas emissions by 50 percent by 2030 ▪ Increase access to renewable energy for people of color and Indigenous communities with community solar and subsidies
Transportation	▪ Support charging for car sharing and extend service to underserved communities
Green Jobs	▪ Partner with frontline communities to develop pathways for living-wage green jobs ▪ Develop a green workforce that is representative of the diversity of the county ▪ Develop job-training programs targeting residents of frontline communities

For example, the Health and Equity section identifies reporting metrics and provides a qualitative assessment of how King County programs are making progress on equity goals (271). It also details its equity accomplishments since 2015, including a 2019 convening of the King County Metro Mobility Equity Cabinet to identify investments needed to provide additional transit service to areas with unmet needs.

If just viewing the climate action plan, it would appear that equity was not a concern in Seattle's climate action planning. This analysis of related plans reveals that Seattle has incorporated equity and justice issues into most areas of its climate planning efforts. King County's 2020 Strategic Climate Action Plan illustrates the comprehensiveness of Seattle's commitment to equity across all aspects of the city's climate agenda. Seattle is clearly a pioneer in equity-focused sustainable and resilient urban planning.

From Planning to Implementation: Where Equity Can Get Sidelined

This review of planning documents in Boston and Seattle highlights the importance of looking beyond climate action plans to get a full view of the city's commitment to social justice and equity within climate action. In both cities, separate planning documents outlined the equity agenda, along with transportation, equity goals, and resilience. Additionally, other climate-related plans identified other equity goals to varying degrees. Both Boston and Seattle separated equity from their main climate action plans. Rather than integration, they chose to write specific equity-focused reports that complemented the main climate action plans. This approach has not necessarily led to more integrated and effective implementation. To assess how well a city is incorporating equity into its climate change planning initiatives, holistic evaluation of all of a city's climate-related plans is necessary.

While urban planning can provide goals and intention to link climate action with social justice, implementation is where tensions and competing priorities thwart transformative change toward equity and social justice. We highlight three barriers evident in one or both cities.

CONFLICTING PRIORITIES

Both Boston's and Seattle's climate and equity plans call for more energy-efficient low-income housing, but neither climate plans address how to increase affordable housing. While there are programs in place, Boston's efforts to date do not provide sufficient affordable housing, let alone affordable and energy-efficient housing. Boston's primary strategy for providing affordable housing, the Inclusionary Development Policy (IDP), has been in place since 2000. It requires developers who seek zoning relief to build and/or finance 13 percent of the units in a development for "income-restricted" tenants or purchasers, or to contribute to an affordable housing fund. This policy has not resulted in sufficient low-income housing providing housing more for moderate and middle-income households. In addition, Boston's comprehensive plan, Imagine Boston 2030 (Boston Planning and Development Agency 2017), sets a goal of seventy thousand income-restricted units by 2030 but doesn't identify how the goal will be achieved.

Seattle also struggles with addressing housing affordability. To expand affordable housing, Seattle voters have passed one bond and five levies since 1981, funding more than thirteen thousand affordable apartments. Still, Seattle ranks in the top twenty unaffordable housing cities. In 2019 the city passed mandatory housing affordability legislation that changes zoning rules to allow denser construction and taller buildings in twenty-seven transit-rich neighborhoods. It also requires developers to maintain between 5 and 11 percent of space in new buildings for low-income households or to make payments into the city's affordable-housing fund (Beekman 2019). Even with this step in the right direction, the policy is only expected to build three to six thousand new houses in the next decade. The point is that in both cities, affordability is a seemingly intractable problem. And given that a city's climate action plans are dependent on other city and state programs, it is understandable that both Boston and Seattle moved to make their climate action plans solely about specific mitigation, adaptation, and resilience goals and to create a separate equity plan that can more easily be integrated with other city efforts.

MISALIGNMENT OF GOALS

Another challenge of implementation is misalignment of priorities, that is, goals of one plan can be misaligned with other city or state goals, policies, or programs. If equity goals do not align with other current policies, they are unlikely to be achieved. We see this in Boston's climate equity planning documents in the context of energy efficiency programs. The top three actions on buildings in Boston's climate action plan are (1) maximize energy efficiency (deep energy retrofits); (2) switch out fossil-fuel space and hot-water heating with GHG-free electricity; and (3) establish strong energy performance standards for all buildings. Carbon Free Boston's equity report revised these approaches from an equity lens, suggesting that the city (1) expand energy-efficient, affordable housing to slow the displacement of vulnerable Bostonians; (2) ensure that the cost of fuel and electricity is not a burden to low-income households; and (3) share quality of life enhancements from energy-efficient housing across the city.

Transitioning from doing as many retrofits as quickly as possible to prioritizing low-income households requires refocusing state programs. In particular, reducing the energy burden requires better integration of policies and programs at the state and municipal levels. Residential energy efficiency programs often focus on market-based solutions, including low-interest loans and tax rebates that are inaccessible to low-income and minority residents who cannot qualify for the loans because of low credit scores and who do not earn enough income to qualify for tax rebates (Reames 2016).

Massachusetts does have several energy-efficiency programs in place that target low-income households. The Low-Income Home Energy Assistance Program (LIHEAP) provides low-income households a fixed benefit amount to help with the cost of their primary heat sources. Mass Save provides free energy efficiency assessments to Massachusetts residents. Low-income residents can qualify for discounted or free energy-efficiency services, including insulation, LED light bulbs, and energy-efficient appliances. Despite the availability of these programs, several factors make it difficult for low-income residents to participate in energy-efficiency programs. Still, the Mass Save energy-efficiency initiative does not adequately serve moderate-income and non–English-speaking households that pay proportionally more for energy-efficiency services (Hasz 2018). This problem is evident in many cities.[2] New funding streams and/or public investments are needed to focus on low-income residents if cities are to promote social justice and equity in energy-efficient housing goals.

DEPARTMENTAL AND BUDGETARY CONTROL

Budgetary control is another key challenge in implementation. Both Boston and Seattle emphasize the need for improved transit accessibility in low-income neighborhoods as part of their climate justice transportation plans. Boston and Seattle both discuss how inequitable the current state of public transit is in their respective transportation plans. Go Boston 2030 (Boston Transportation Department 2017) plans to expand access for low-income areas by making every home in Boston within a ten-minute walk of a rail station or a major bus route. It also has set the 2030 reliability goal of decreasing commute times by 10 percent. Seattle's City Light Transportation Electrification Strategic Investment Plan (Seattle City Light, 2019) fell into the "Integrated" category in our analysis because of its inclusion of equity and social justice in many areas of its transit planning initiatives. For example, Seattle is electrifying its bus fleet to decrease pollution in low-income areas while also expanding its bus service.

Despite having these transit equity goals, neither Boston nor Seattle operates its transit agency. Boston is served by the regional Massachusetts Bay Transit Authority (MBTA and Seattle by the Central Puget Sound Regional Transit Authority (Sound Transit). Both agencies, like many in the country, are fiscally stressed, which was made worse by the COVID-19 pandemic, which is making proposed expansions in access and transit equity goals even more difficult.

Conclusion

Although equity already has been incorporated in climate action in Boston and Seattle, we note that racial equity is seldom mentioned specifically. That is beginning to change in several cities. For example, Austin, Texas, started updating its climate plan in the months before the COVID-19 lockdown began, with an explicit mandate to incorporate racial justice as an overarching theme. This emphasis is reflected in its title—the Austin Climate Equity Plan. Adopted in September 2020, the ambitious plan begins with an acknowledgment of the city's racist history and promises to address disproportionate environmental harm done historically and currently in low-income communities and communities of color. The pandemic has revealed and exacerbated economic inequities and racial disparities in cities, creating a renewed urgency for climate action planning to advance racial and economic justice. Austin's plan stands as an example for other cities to follow.

It is useful to analyze planning documents with respect to incorporation of equity goals. The proof of a plan, however, is in implementation. Many an idealistic agenda developed by urban planners gets sidelined or undermined by politicians, developers, or private-sector interests with a competing agenda. We see this particularly in cities with hot development and housing markets such as Boston and Seattle. It is no secret that developers wield immense power over the politics of most cities and states. When it comes to urgent climate policy, their short-term interests must not be allowed to call the tune. And city officials will almost always choose economic development concerns over environmental and/or equity concerns. The environmental justice and community development organizations involved in developing city climate justice plans will need to apply ongoing pressure to ensure that racial and equity and social justice goals are not compromised.

We are in a new postpandemic era. The Biden administration has demonstrated a commitment to climate action and to racial and economic justice that will create new opportunities for cities to leverage national-level climate programs. Let us hope that that this moment is not, in the words of British historian G. M. Trevelyan describing another time, a turning point at which modern history failed to turn.

1. The CDP rankings can be viewed at https://www.cdp.net/en/cities/cities-scores.

2. In New York City, inequities in access to urban renewable energy and energy-efficiency programs stem from a combination of gaps in energy literacy, lack of prioritization of energy efficiency among other pressing issues, limited access to participation in the political process and discrimination in opportunities to invest in energy-efficient infrastructure (City of New York, 2019).

REFERENCES

Agyeman, J., Schlosberg, D., Craven, L., and Matthews, C. (2016). Trends and directions in environmental justice: From inequity to everyday life, community, and just sustainabilities. *Annual Review of Environment and Resources*, 41, 321–340.

Anguelovski, I., Connolly, J., and Brand, A. L. (2018). From landscapes of utopia to the margins of the green urban life: For whom is the new green city?. *City*, 22(3), 417–436.

Balk, G. (2017). Seattle hits record high for income inequality, now rivals San Francisco. *Seattle Times*, November 17. https://www.seattletimes.com/seattle-news/data/seattle-hits-record-high-for-income-inequality-now-rivals-san-francisco/.

Beekman, D. (2019). Seattle's upzones would collect millions for affordable housing. Who would get help? *Seattle Times*, March 14. https://www.seattletimes.com/seattle-news/politics/seattles-upzones-would-direct-millions-to-affordable-housing-what-does-that-look-like/.

Béné, C., Mehta, L., Mcgranahan, G., Cannon, T., Gupte, J., and Tanner, T. (2018). Resilience as a policy narrative: Potentials and limits in the context of urban planning. *Climate and Development*, 10(2), 116–133.

Berube, A. (2018). *City and Metropolitan Income Inequality Data Reveal Ups and Downs through 2016*. Brookings.

Boston Green Ribbon Commission. (2019). Carbon Free Boston: Social Equity Report. https://www.greenribboncommission.org/wp-content/uploads/2019/05/CFB_Social_Equity_Report_WEB.pdf.

Boston Housing Authority. (2014). Strategic Sustainability Plan. https://www.bostonhousing.org/BHA/media/Documents/BHA-Sustainability_Feb18.pdf.

Boston Planning and Development Agency. (2017). Imagine Boston 2030 Implementation. http://www.bostonplans.org/planning/imagine-boston-2030-implementation.

Boston Transportation Department. (2017). Go Boston 2030 Vision and Action Plan. https://www.boston.gov/sites/default/files/file/document_files/2019/06/go_boston_2030_-_full_report.pdf.

Bulkeley, H., Carmin, J., Castán Broto, V., Edwards, G. A. S., and Fuller, S. (2013). Climate justice and global cities: Mapping the emerging discourses. *Global Environmental Change*, 23(5), 914–925.

Bullard, R. D. (1994). Overcoming racism in environmental decision-making. *Environment: Science and Policy for Sustainable Development*, 36(4), 10–44.

Carbon Neutral Cities Alliance (CNCA). (2020). Seattle, Washington, USA. https://carbonneutralcities.org/cities/seattle/.

Carley, S., and Konisky, D. M. (2020). The justice and equity implications of the clean energy transition. *Nature Energy*, 5, 569–577.

Chen, J. T., and Krieger, N. (2020). Revealing the unequal burden of COVID19 by income, race/ethnicity, and household crowding: U.S. county versus zip code analyses. *Journal of Public Health Management and Practice*, Jan/Feb 27 Suppl., COVID-19 and Public Health: Looking Back, Moving Forward, S43–S56.

City of Boston. (2017). Economic Inclusion + Equity Agenda. https://www.cityofboston .gov/pdfs/economicequityinclusionagenda.pdf.

———. (2019a). Climate Vulnerability Assessment. https://www.boston.gov/sites/default /files/imce-uploads/2017-01/crb_-_focus_area_va.pdf.

———. (2019b). Housing a Changing City Boston 2030: 2018 Update, Q1 Progress Report. https://docs.google.com/document/d/1QcvaAmLBQv4K6QXagVmN1CHj4Lg _ryH_AT6epZtOIhQ/edit.

———. (2019c). Mayor's Food Access Agenda, 2019–2021. https://www.boston.gov/sites /default/files/imce-uploads/2019-09/mayors_food_acees_agenda_external_final _9_11_19_.pdf.

City of New York. (2019). OneNYC 2050. https://onenyc.cityofnewyork.us.

City of Seattle. (2018). Seattle Climate Action. http://greenspace.seattle.gov/wp-content /uploads/2018/04/SeaClimateAction_April2018.pdf.

City of Seattle. (2020). Seattle Housing Levy. https://www.seattle.gov/housing/levy.

Climate Resilience Initiatives. (2019). https://www.boston.gov/departments/environment/preparing-climate-change (no longer available).

Commonwealth of Massachusetts. (2020). Environmental Justice Populations in Massachusetts. https://www.mass.gov/info-details/environmental-justice-populations-in -massachusetts.

Daniels, L., and O'Donnell, B. (2019). Seattle City Light Transportation Electrification Strategy. Rocky Mountain Institute. https://powerlines.seattle.gov/wp-content /uploads/2019/10/City-Light-Transportation-Electrification-Strategy.pdf.

Esri.com. (2017). Income Inequality in Boston. https://storymaps.esri.com/stories/2015 /map-journal-smaller-embed/?appid=5e7116880ec34270bc88c00c41dbda6d.

Fainstein, S. S. (2018). Resilience and justice: Planning for New York City. *Urban Geography*, 39(8), 1268–1275.

Finn, D., and McCormick, L. (2011). Urban climate change plans: How holistic?. *Local Environment*, 16(4), 397–416.

Fitzgerald, J. (2020). *Greenovation: Urban Leadership on Climate Change*. Oxford University Press.

———. (2022). Transitioning from urban climate action to climate equity. *Journal of the American Planning Association*, https://www.tandfonline.com/doi/full/10.1080/01944 363.2021.2013301.

Fitzgibbons, J., and Mitchell, C. L. (2019). Just urban futures? Exploring equity in "100 Resilient Cities." *World Development*, 122, 648–659.

Foster, J., Uhlig, A. D., Cutler, D., Reed, A., Wong, T., and Hamilton, L. (2013). Seattle Climate Action Plan. http://www.seattle.gov/Documents/Departments/Environment /ClimateChange/2013_CAP_20130612.pdf.

Greenovate Boston. (2014). Climate Action Plan Update. https://www.cityofboston.gov /eeos/pdfs/Greenovate%20Boston%202014%20CAP%20Update_Full.pdf.

Guyadeen, D., and Seasons, M. (2016). Plan evaluation: Challenges and directions for future research. *Planning Practice and Research*, 31(2), 215–228.

Harris, L. M., Chu, E. K., and Ziervogel, G. (2017). Negotiated resilience. *Resilience*, 6(3), 196–214.

Hasz, A. (2018). *Equitable Energy for Massachusetts: How Can Climate Policy Reduce Inequality?* Massachusetts Institute of Technology, master's thesis.

Hess, D. J., and Satcher, L. A., (2019). Conditions for successful environmental justice mobilizations: An analysis of 50 cases. *Environmental Politics*, 28(4), 663–684.

Hightower, K. (2020). Mayor Durkan announces bold action to deliver 500 permanent homes in response to COVID-19 public health crisis. Citywide Strategy, Seattle Office of the Mayor. https://durkan.seattle.gov/2020/06/mayor-durkan-announces-bold -action-to-deliver-500-permanent-homes-in-response-to-covid-19-public-health -crisis/.

Hughes, A., and Hoffmann, M. (2020). Just urban transitions: Toward a research agenda. *Wiley Interdisciplinary Reviews: Climate Change*, 11(3), E640.

Junejo, S. (2017). Equity Analysis of Washington State Toxics Sites and the Model Toxic Control Act. Front and Centered, Seattle. https://frontandcentered.org/wp-content /uploads/2017/01/MTCA-Report_1-25-17.pdf.

Kaika, M. (2017). "Don't call me resilient again!": The new urban agenda as immunology . . . or . . . what happens when communities refuse to be vaccinated with "smart cities" and indicators. *Environment and Urbanization*, 29(1), 89–102.

King County, Washington. (2020). Strategic Climate Action Plan. https://kingcounty .gov/services/environment/climate/actions-strategies/strategic-climate-action-plan .aspx.

Landry, F., Dupras, J., and Messier, C. (2020). Convergence of urban forest and socioeconomic indicators of resilience: A study of environmental inequality in four major cities in Eastern Canada. *Landscape and Urban Planning*, 202, 103856.

Lukanov, B. R., and Krieger, E. M. (2019). Distributed solar and environmental justice: Exploring the demographic and socio-economic trends of residential PV adoption in California. *Energy Policy*, 134, 110935.

Meerow, S., Pajouhesh, P., and Miller, T. R. (2019). Social equity in urban resilience planning. *Local Environment*, 24(9), 793–808.

Meerow, S., and Newell, J. P. (2017). Spatial planning for multifunctional green infrastructure: Growing resilience in Detroit. *Landscape and Urban Planning*, 159, 62–75.

Mohai, P., and Saha, R. (2015). Which came first, people or pollution? Assessing the disparate siting and post-siting demographic change hypotheses of environmental injustice. *Environmental Research Letters*, 10(11), 115008.

Muro, M., Tomer, A., Shivaram, R., and Kane, J. (2019). Advancing Inclusion through Clean Energy Jobs. Metropolitan Policy Program. Brookings. https://www.brookings.edu/wp-content/uploads/2019/04/2019.04_metro_Clean-Energy-Jobs_Report_ Muro-Tomer-Shivaran-Kane.pdf.

Pastor, M., and Morello-Frosch, R. (2018). Gaps matter: Environment, health, and social equity. *Generations*, 42(2), 28–33.

Patton, M. Q. (2002). Utilization-Focused Evaluation Checklist. https://wmich.edu/sites/ default/files/attachments/u350/2014/UFE_checklist_2013.pdf.

Reames, T. G. (2016). Targeting energy justice: Exploring spatial, racial/ethnic and socioeconomic disparities in urban residential heating energy efficiency. *Energy Policy*, 97, 549–558.

Reames, T. G., Reiner, M. A., and Stacey, M. B. (2018). An incandescent truth: Disparities in energy-efficient lighting availability and prices in an urban U.S. county. *Applied Energy*, 218, 95–103.

Ribeiro, D., Samarripas, S., Tanabe, K., Jarrah, A., Bastian, H., Drehobl, A., et al. The 2020 City Clean Energy Scorecard. American Council for an Energy Efficient Economy, Washington, D.C.

Samarripas, S., Tanabe, K., Dewey, A., Jarrah, A., Jennings, B., Drehobl, A., et al. (2021). The 2021 City Clean Energy Scorecard. American Council for an Energy-Efficient Economy, Washington, D.C. aceee.org/research-report/u2107.

Schrock, G., Bassett, E. M., and Green, J. (2015). Pursuing equity and justice in a changing climate: Assessing equity in local climate and sustainability plans in U.S. cities. *Journal of Planning Education and Research*, 35(3), 282–295.

Seattle City Light. (2019). Transportation Electrification Strategic Investment Plan. https://www.seattle.gov/documents/Departments/CityLight/TESIP.pdf.

Seattle Office of Sustainability and Environment. (2017). Equity and Environment Agenda. https://www.seattle.gov/Documents/Departments/OSE/Equity/SeattleEE-Agenda.pdf.

Shonkoff, S. B., Morello-Frosch, R., Pastor, M., and Sadd, J. (2011). The climate gap: Environmental health and equity implications of climate change and mitigation policies in California—a review of the literature. *Climatic Change*, 109(S1), 485–503.

Sovacool, B. K. (2014). What are we doing here? Analyzing fifteen years of energy scholarship and proposing a social science research agenda. *Energy Research and Social Science*, 1, 1–29.

Sovacool, B. K., Burke, M., Baker, L., Kotikalapudi, C. K., and Wlokas, H. (2017). New frontiers and conceptual frameworks for *Energy Justice. Energy Policy*, 105, 677–691.

Steele, W., Maccallum, D., Byrne, J., and Houston, D. (2012). Planning the climate-just city. *International Planning Studies*, 17(1), 67–83.

Stephens, J. C. (2020). *Diversifying Power: Why We Need Antiracist, Feminist Leadership on Climate and Energy*. Island Press.

Tessum, C. W., Paolella, D. A., Chamblis, S. E., Apte, J. S., Hill, J. D., and Marshall, J. D. (2021). Polluters disproportionately and systemically affect people of color in the United States. *Science Advances*, 7(18), https://www.science.org/doi/10.1126/sciadv.abf4491.

Warner, M. E. (2010). The future of local government: Twenty-first-century challenges. *Public Administration Review*, 70, S145–S147.

Waud, A. (2017). Cities, Social Justice, and Striving for Carbon Neutrality. Unpublished paper, University of Toronto.

York, D., and Jarrah, A. (2020). Community Resilience Planning and Clean Energy Initiatives: A Review of City-Led Efforts for Energy Efficiency and Renewable Energy. American Council for an Energy-Efficient Economy, Washington, D.C.

Ziervogel, G., Pelling, M., Cartwright, A., Chu, E., Deshpande, T., Harris, L., et al. (2017). Inserting rights and justice into urban resilience: A focus on everyday risk. *Environment and Urbanization*, 29(1), 123–138.

CHAPTER 6

Bringing Equity into Climate Change
Adaptation Planning in New York City

ROBIN LEICHENKO, SHEILA R. FOSTER,
AND KHAI HOAN NGUYEN

For cities within the United States, efforts to promote climate justice are inextricably tied to the growing awareness of practices and policies such as racialized policing and redlining that contribute to and perpetuate structural racism (Sealey-Huggins 2018; Rice, Levenda, and Long, this volume). George Floyd's cry, "I can't breathe," not only highlighted the shocking reality of police brutality in communities of color but also sparked renewed attention to the ways that systemic racism also makes it difficult for those communities to, collectively, breathe. Communities of color suffer disproportionately from asthma, respiratory illness, heart disease, and COVID-19 infection (Gaynor and Wilson 2020; Abedi et al. 2021). These same communities are often highly exposed to air pollution and are among the most vulnerable to the impacts of climate change such as flooding and heat waves. This heightened recognition of how inequities can be structurally embedded into urban landscapes is especially relevant for urban climate change adaptation efforts, where efforts to create resilience to heat waves, flooding, wildfires, and droughts may necessitate long-term shifts in land-use practices and changes in spatial configurations and functioning of cities.

Growing awareness of the role of systemic racism in shaping communities' vulnerability to climate change is paving the way for new kinds of "praxis" in the way that some cities approach adaptation planning in thick urban environments, that is, urban environments with high levels of diversity and significant social and economic inequalities between communities. This chapter is a reflection on the authors' role on the New York City Panel on Climate Change (NPCC), and specifically our role leading its workgroup on Community-Based Assessment of Adaptation and Equity (CBA-Equity) (Foster et al. 2019). The NPCC is an independent advisory body that is charged in local New York

City law with providing the City of New York with regularly updated climate change scenarios and information on climate change impacts (Solecki 2012).

The authors (Foster and Leichenko) were appointed to the NPCC by New York City's mayor based on their expertise on climate change and environmental justice and their ability to offer scientific advice to the city in order to help it prepare for the impacts of climate change. This was the first time since its inception that the mayor of New York City instructed the NPCC to address climate change impacts and adaptation through the lens of equity and at the neighborhood level, with explicit focus on community-based adaptation. While prior New York City and New York State work had identified a need for consideration of equity and environmental justice in the analysis of climate impacts, vulnerabilities, and adaptation (NYCEJA 2018; NYCEJA 2016; Sandy Regional Assembly 2013; Leichenko et al. 2011), the formation of the CBA Equity Workgroup within the NPCC reflected the city's heightened recognition of and desire for action on these issues (Foster et al. 2019).

This chapter describes our workgroup's efforts to identify ways to incorporate equity into the city's climate change vulnerability analyses and adaptation planning efforts (Foster et al. 2019). Each of the chapter authors are academics—Leichenko is a professor of geography at Rutgers University, Foster is a professor of law and public policy at Georgetown University, and Nguyen is a PhD student in geography at Rutgers. The overall composition of the workgroup consisted of six academic researchers, four graduate students (including Nguyen), and three representatives of local, grassroots environmental justice (EJ) organizations. The racial makeup of the workgroup group was approximately 60 percent white and 40 percent nonwhite.

The workgroup's approach entailed a coproduction model that was designed to meaningfully engage local communities to collaboratively identify key climate vulnerabilities and related stresses and to assess how best to incorporate equity into adaptation planning (Deas et al. 2017; Sarzynski 2015; Leichenko et al. 2014; Lemos and Morehouse 2005; Cole and Foster 2001). The workgroup met at the outset with representatives from the city and from three local community-based organizations (CBOs) representing "frontline" communities—WE ACT for environmental justice in Harlem, THE POINT CDC in Hunts Point, and UPROSE in Sunset Park. The CBOs were included as full participating members and contributors to the research, helping ensure that the focus of the work was guided by community concerns and that the process and product adhere to the inclusive principles of environmental justice (see Foster et al. 2019 for a full description of the work).

While environmental justice activism in New York City can be traced back decades, it was only recently recognized in the city's climate adaptation efforts. A key initial challenge for the workgroup was navigating competing understandings of the meaning of equity in adaptation planning. From the perspective of the city, the primary initial interests for the work included guidance on the use of social vulnerability mapping as a tool for identifying spatial vulnerability patterns and areas in need of resources and case studies of socially vulnerable communities. By contrast, the CBOs were more interested in having a voice in adaptation decision-making within their neighborhoods and viewed participation in the workgroup as a means of having a voice. The CBOs also wanted the workgroup to pay explicit attention to the inequities manifesting in their communities, particularly economic processes facilitating gentrification and structural racism. As described in the next sections, the workgroup embraced these competing definitions of equity through adoption of a three-part equity framework. The workgroup implemented this framework, first through collaborative exploration of social vulnerability maps, which emphasized distributive equity, and then through coproduction of community case studies that focused on procedural and contextual equity. The workgroup's efforts highlight the vital importance of community partnerships in all phases of urban climate adaptation planning, reinforcing the foundational need for attention to equity and justice as means to secure urban futures (Rice, Levenda, and Long, this volume).

An Equity Framework for Climate Change Adaptation Assessment in New York

By adopting a coproduction model, members of the workgroup were putting to the test what "meaningful engagement" could look like as means to integrate equity and environmental justice into the city's larger climate action agenda. A key initial challenge for the workgroup was the establishment of a sense of trust of the NPCC process with the CBOs. The CBOs have each at times contested that city actions and policies have affected environmental quality in their community (e.g., siting of noxious facilities). Each was initially somewhat skeptical about whether collaboration with the NPCC would be meaningful and productive for their organization, given the history of mixed outcomes in their engagement with the city around environmental justice issues.

In our view, three factors likely contributed to establishment of trust between the CBO representatives and the NPCC. One factor was Foster's record of environmental justice scholarship and her long history of work with these and other EJ CBOs in New York City. Her reputation and established relationship with the leaders of two of the CBOs were especially critical for initiating the NPCC collaboration. Another important factor was transparency about the role and ability of the NPCC to influence city action. In discussing the influence that our work might have on New York City policy, we advised the groups that their participation would be a way for environmental justice organizations to have a voice in guiding the NPCC on how the city can best incorporate equity into adaptation planning. While there was no way to guarantee that the city would use this information as recommended, the fact that the city solicited guidance on equity in adaptation was viewed as important and relevant to the mission of the CBOs. A third factor was the collaborative and coproduction-based approach of the work and our explicit commitment to principles of environmental justice in how our workgroup operated. In particular, the EJ groups would have the leading voice in the identification of key climate risks for their communities as well as adaptation needs. The groups also had an important collaborative role in the framing of the report's findings and its recommendations about how to make the city's adaptation planning process more inclusive and equitable.

To make sense of the differing understandings of equity between the city and environmental groups—namely, distributive versus procedural equity—the workgroup drew on environmental justice praxis and literatures, as well as recent equity-focused contributions within the climate change adaptation and mitigation literatures (e.g., Meerow, Pajouhesh, and Miller 2019; Foster 2017; Schlosberg and Collins 2014; McDermott, Mahanty, and Schreckenberg 2013; Leichenko et al. 2011; Cole and Foster 2001). The workgroup also drew on McDermott, Mahanty, and Schreckenberg's (2013) concept of contextual equity, which focuses on economic and social processes that contribute to marginalization. The workgroup ultimately adopted the three-part equity framework developed by McDermott, Mahanty, and Schreckenberg (2013), with modifications to reflect our focus on urban adaptation (see table 6.1).

Within the framework, distributive equity emphasizes the uneven environmental burdens and benefits across groups and neighborhoods (Foster 1998). This interpretation reflects the suggestions of the environmental justice groups as well as the literature on environmental justice, which has brought attention to racial and ethnic disparities in the distribution of polluting facilities and other environmental hazards and the lack of environmental amenities such as green and open spaces in low-income and minority communities (Corburn, Osleeb, and Porter 2006; Cole and Foster 2001; Fothergill, Maestas, and Darlington 1999; U.S. EPA 1992). This approach also incorporates more recent climate change literature, where elements of distributive equity include recog-

TABLE 6.1. Three-dimensional equity framework (based on McDermott, Mahanty, and Schreckenberg 2013)

Distributive equity	Emphasizes disparities across social groups, neighborhoods, and communities in vulnerability, adaptive capacity, and the potential for socially and spatially uneven outcomes of adaptation actions
Contextual equity	Emphasizes social, economic, and political factors and processes, including systemic and structural racism, that contribute to uneven vulnerability and shape adaptive capacity
Procedural equity	Emphasizes the extent and robustness of public and community participation in adaptation planning and decision-making

nition of inequalities in social vulnerability to climate change; inequalities in the capacity to adapt or influence mitigation of climate change; inequalities in benefits associated with adaptation policies; and inequalities and unintended consequences of adaptation and mitigation efforts (McDermott, Mahanty, and Schreckenberg 2013; Leichenko et al. 2011). Both sets of literatures bring attention to the distribution of costs and benefits of policy initiatives on various populations. Rooted in principles of equality and social welfare, these approaches are often needs-based and directly target the least advantaged communities and the most at-risk community members in standard-setting and adaptation planning (McDermott, Mahanty, and Schreckenberg 2013).

Contextual equity is a relatively recent addition to the climate change literature (McDermott, Mahanty, and Schreckenberg 2013). However, its essential elements are well recognized in the climate vulnerability and environmental justice literatures, both of which emphasize social "root causes" of existing disparities and vulnerabilities, including the influence of social context and structural racism (Ribot 2014; Cole and Foster 2001; Sarzynski 2015). Within our framework, contextual equity draws attention to factors that contribute to social vulnerabilities and recognizes that differences in power and access can prevent some communities from receiving resources or from participating in the decision-making process (Fraser 2009). Acknowledging the "uneven playing field" that is created for some communities as a result of preexisting economic, social, and political inequalities (McDermott, Mahanty, and Schreckenberg 2013), contextual equity draws attention to socioeconomic conditions and existing injustices that are critical for designing community-based adaptation strategies (Schlosberg, Collins, and Niemeyer 2017).

Procedural equity is typically defined as the representation and inclusion of affected individuals, communities, and groups in environmental and adaptation priority-setting and decision-making. With respect to climate change, this includes decisions about adaptation strategies and actions, as well as emergency preparedness and emergency response in relation to climate-related risks. Efforts to achieve procedural equity often require explicit mechanisms to ensure participation of affected actors in policy and planning decisions (Chu, Anguelovski, and Carmin 2016; Schlosberg 2013; Leichenko et al. 2011). Traditional efforts to include groups historically deprived of resources in environmental and adaptation decision-making processes include public hearings and meetings, citizen advisory councils, and citizen panels (Sarzynski 2015). However, the climate change community is also paying increased attention to the need for greater inclusion of affected groups in the climate assessment process, including identification of critical risks and vulnerabilities, formulation of adaptation options, and selection and implementation of response strategies (Cornell et al. 2013; Kirchhoff, Lemos, and Dessai 2013; Rosenzweig et al. 2011). This type of collaborative engagement of affected communities in all phases of adaptation planning and implementation has been identified by the environmental justice community as a critical need in the New York region (NYCEJA 2018; NYCEJA 2016; Sandy Regional Assembly 2013).

The workgroup functioned via a collaborative approach where CBO representatives were engaged in all phases of the research. Members of the workgroup initially met in person with leaders of each of the CBOs and asked them to join the workgroup efforts. Each of the CBOs had a history of successful environmental justice activism in New York City, and each was already deeply engaged in climate adaptation, mitigation, and resilience projects. As a first step for the work, CBO representatives collaborated with the workgroup members to identify climate risks, vulnerabilities, and related stressors in order to gain a better and more complete picture of distributive and contextual equity concerns in the three communities. Because each of the CBOs was already engaged in community-based climate resilience efforts, their representatives were readily able to pinpoint key risks and vulnerabilities within each local community. Frank discussions with CBO leaders also provided important insights into their interactions with the city's climate mitigation and adaptation efforts as a lens into the issue of procedural equity. In addition to collaboration with representatives from each of the CBOs, the workgroup also interviewed city officials, reviewed policy and planning documents from both the city and the CBOs, and collected relevant demographic and health data from city agencies and public sources. CBO representatives provided feedback and comments on draft versions of the workgroup report, which were incorporated into the final version (Foster et al. 2019).

We drew on the equity framework for all phases of the work. Distributive equity was central for our examination of spatial patterns of vulnerability to climate change stresses across neighborhoods and our recommendations on methods and indicators for monitoring and tracking neighborhood vulnerability. Contextual equity was featured in our case studies of community vulnerability and adaptation. Procedural equity was incorporated into our assessment of how community groups are included in the development and implementation of adaptation plans. Each of these phases of the study is described briefly in subsequent sections. Full results are presented in Foster et al. (2019).

Distributive Equity:
Vulnerability Mapping and Targeting of Resources

Consideration of distributive equity is foundational for social vulnerability mapping and analysis where the goal of the work is documentation of uneven distribution of vulnerabilities to climate shocks and stress across neighborhoods, communities, and regions (Cutter and Finch 2008; Adger 2006; Cutter, Mitchell, and Scott 2000). In addition to measuring vulnerability to climate stressors, social vulnerability analysis also is widely used to measure exposure to toxic and hazardous facility siting and to determine "environmental justice" areas based on indicators that track proximity to a variety of pollution sources (Foster 2017; Sadd et al. 2011). Factors that are often found in both literatures include socioeconomic status (wealth or poverty); education; age; access and functional needs; gender; race and ethnicity (Cutter et al. 2009) (see Foster et al. 2019 for a complete review). Through the creation of empirical metrics and indicators of social vulnerability, researchers capture a wide array of factors that shape the susceptibility of certain populations and communities to harm from environmental hazard events and the ability to recover following these events (Tate 2012; Cutter, Boruff, and Shirley 2003). These analyses are often explicitly designed to help identify "hot spots" for needs-based targeting of resources and policies to communities that are most at risk (de Sherbinin 2014; Dunning and Durden 2011).

In exploring options for documenting and tracking spatial vulnerability in New York City, our workgroup explored a variety of methodological approaches used for social vulnerability analysis and mapping in New York and elsewhere. These include mapping applications conducted by nonprofit organizations, academic institutions, and governmental agencies (HVRI 2018; CDC SVI 2018). Among these studies, we identified two common and widely used approaches for vulnerability mapping. These include the SoVI, developed by Susan Cutter (Cutter, Boruff, and Shirley 2003), and the Social Vulnerabil-

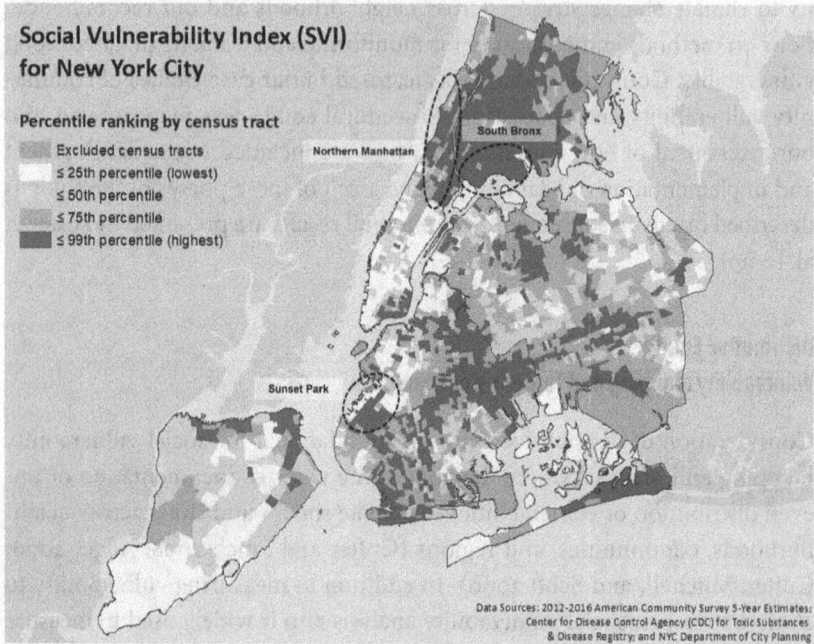

Social Vulnerability Index (SVI) for New York City

Percentile ranking by census tract
- Excluded census tracts
- ≤ 25th percentile (lowest)
- ≤ 50th percentile
- ≤ 75th percentile
- ≤ 99th percentile (highest)

Northern Manhattan

South Bronx

Sunset Park

Data Sources: 2012-2016 American Community Survey 5-Year Estimates; Center for Disease Control Agency (CDC) for Toxic Substances & Disease Registry; and NYC Department of City Planning

FIGURE 6.1. Social Vulnerability Index (SVI) results for New York City (based on New York State calculation).

ity Index (SVI), developed by the CDC (Flanagan et al. 2011). Both methods have been empirically validated and replicated and are widely used throughout the United States (Reckien 2018; Bakkensen et al. 2017; Myers, Slack, and Singelmann 2008; Cutter and Emrich 2006). For New York City, creation of social vulnerability maps based on either method could aid in the identification of census tracts with high levels of social vulnerability to all types of climate stressors including heat, floods, and other types of stressors.

As illustrated in figure 6.1, which depicts a vulnerability map of New York City created using the SVI method, these maps provide information on spatial patterns across different communities or neighborhoods. The mapping results reveal that social vulnerability is unequally distributed across New York City; high levels of social vulnerability are consistently found in areas with lower incomes and higher shares of African American and Hispanic residents. Our three case study communities, discussed in the next section, are identified as having high levels of social vulnerability.

Despite the widespread usage of social vulnerability analysis, the workgroup also noted several limitations of vulnerability indices for application to policy and planning decisions (Preston, Yuen, and Westaway 2011; Schmidtlein et al. 2008). For example, social vulnerability scores, which are employed to map and visualize patterns of social vulnerability, only provide a relative indicator of vulnerability in comparison to other areas. In other words, a low vulnerability score simply means that one area has relatively lower social vulnerability than areas with higher scores; a low vulnerability score does not ensure that an area is resilient to climate shocks, nor does it imply that all residents of that area have low vulnerability. This type of aggregated, composite vulnerability index has more limited utility for tracking how vulnerability changes over time in a particular community or geographic area. The numerical score values for individual tracts are not directly comparable over time because the scores for each period are calculated relative to other tracts during that same period. In addition, the scores do not provide clear guidance on which components of social vulnerability have contributed to changes in score values. For these reasons, tracking changes in social vulnerability over time can be better accomplished using single variable indicators.

As an alternative or supplement to construction of vulnerability indices, the workgroup recommended consideration tracking of specific indicators on neighborhood vulnerability over time. Use of specific indicators would permit documentation of changes over time and ensure continual needs-based targeting of adaptation efforts. While many factors contribute to social vulnerability of specific households or groups, the above approaches permit identification of variables that are widely found to be indicative of social vulnerability (NAACP 2015). The proposed variables (see table 6.2), all of which were found to contribute to social vulnerability in the studies reviewed above, are intended to provide a starting point for vulnerability tracking in New York City. Each variable is readily available from census data sources, and each may be supplemented with additional indicators that are viewed as relevant by the city or by particular communities.

These proposed indicator variables, which are updated annually by the American Community Survey (ACS) at the census tract level, would allow for the tracking of factors that are widely thought to contribute to social vulnerability and spatial differences or inequalities in vulnerability. The indicators are intended to capture demographic, economic, housing, and educational disparities across neighborhoods. They also capture access and functional needs populations and older populations who are especially at risk to climate extremes (Kinney et al. 2015). The workgroup suggested that the tracking process could be supplemented as needed using city health data sources (e.g., NYC Environment and Health Data Portal) to ensure accurate documentation of access and functional needs populations. Additional city-specific

TABLE 6.2. Proposed list of vulnerability indicators for New York City (Foster et al. 2019).

Vulnerability factor	Potential social indicators
Access and functional needs populations	Percent of civilian noninstitutionalized population with a disability
Educational attainment	Percent population with bachelor's degree or higher
	Percent population over 25 years old with no high school degree
English fluency	Percent population 5 years or over who speak English less than "very well"
Female-headed household	Percent of female-headed households
Foreign-born population	Percent of foreign-born population
Income	Median household income
	Percent of households receiving public assistance income
Older adults over 65	Percent population over 65 years old
Poverty	Percent of population living below poverty level
Race/ethnicity	Percent of nonwhite population
Rent burden	Percent of occupied units paying 35 percent or more of household income on rent

health-related variables might include, for example, population lacking air-conditioning, population lacking health insurance, population living with chronic health conditions, population with asthma, and population dependent on electric medical equipment (Kinney et al. 2015; McArdle 2013).

Our workgroup's social vulnerability mapping analysis revealed important information about distributional inequalities in susceptibility to harm as a result of climate change, and how these inequalities vary across New York City communities. While such information can serve as a useful tool for needs-based targeting of adaptation resources, social vulnerability mapping does not illuminate why certain neighborhoods are more vulnerable than others. To effectively address or reduce social vulnerability to climate change, it is necessary to consider the contextual factors that shape the vulnerability of a particular neighborhood or community. We consider these factors in the next section's examination of contextual equity.

Contextual Equity in Socially Vulnerable Communities

A core tenet of the environmental justice movement is that environmentally overburdened communities should "speak for themselves" with regard to the ways in which they suffer the injustice of disproportionate hazard exposure (Bullard and Alston 1990). As such, the very concept of environmental injustice (or inequality) is rooted in the idea of contextual equity. Scholars have articulated and analyzed the theory of environmental injustice "from the ground up," investigating and listening to (as well as capturing the voices of) communities as a window into economic and social factors and dynamics that render those communities vulnerable to disproportionate hazard exposure (Cole and Foster 2001; Foster 1998). Following this approach to the issue of climate justice and equity, the workgroup conducted case studies of three environmental justice communities in New York City: Northern Manhattan, Manhattan; Sunset Park, Brooklyn; and Hunts Point, the Bronx (see figure 6.1) to better understand the interaction between environmental and climate stressors and social and economic disadvantages. The three case studies provided contextual information about these predominantly racial and ethnic minority, low-income communities and the critical climate and nonclimate stressors that affect them. The case studies highlighted many commonalities across the three communities. Communities in Northern Manhattan, Sunset Park, and Hunts Point are each highly vulnerable to climate change based on the vulnerability mapping analysis (see figure 6.1). The three are all also confronting the challenge of gentrification and/or displacement (Austensen et al. 2015). In particular, CBOs identified numerous concerns related to changing social and economic conditions, including, for example, concern about the rising cost of living, increased rents, and lack of affordable housing options (see table 6.3) (Austensen et al. 2015).

In addition, the processes of deindustrialization and commercialization create great uncertainty regarding job opportunities. At the same time, there is an increased presence of commercial development in all three areas, offering unskilled jobs in the service sector (as compared to skilled manufacturing and industrial jobs). These jobs do not allow existing residents to meet increases in the cost of living, particularly housing. The growth of the commercial sector also contributes to conflicts over land use and economic development planning. Vacant warehouses and buildings are being bought by private developers, which threatens to transform working-class neighborhoods into unaffordable upscale enclaves. Residents and community activists are actively fighting to preserve their manufacturing zoning and job opportunities for residents (Fainstein 2018; Sze and Yeampierre 2018; Checker 2011). New commercial activities typically cater to middle- and upper-middle-class clientele and are generally not accessible to low-income residents (Adams 2016; Gonzalez 2016).

The neighborhoods of Northern Manhattan, Sunset Park, and Hunts Point are also considered hotspots of environmental pollution (see table 6.3). They are disproportionately burdened with numerous noxious and polluting industrial facilities and related activities (e.g., garbage processing centers, power

TABLE 6.3. Summary of social, economic, climate, and other environmental stressors and community needs identified by CBOs in the three case study communities (Foster et al. 2019).

Communities	Northern Manhattan	Sunset Park Brooklyn	Hunts Point The Bronx
SOCIAL AND ECONOMIC STRESSORS			
Aging housing stock	X	X	X
Decrease in manufacturing jobs	—	X	—
Energy cost burdens	X	—	—
Health disparities	X	—	X
High share foreign-born residents	X	X	—
High rate of poverty	X	X	X
Increase in commercial presence	X	X	—
Lack of affordable housing options	X	X	X
Rising cost of living	X	X	—
Unemployment	—	—	—
CLIMATE STRESSORS			
Rising average temperatures	X	X	X
Risk in heat waves and hot days	X	X	X
Changing precipitation; inland flooding	X	X	X
Sea level rise	X	X	X
Coastal flooding	X	X	X
Extreme hurricane winds	X	X	X
Drought	—	—	—
Cold snaps	—	—	—
OTHER ENVIRONMENTAL STRESSORS			
Air pollution	X	X	X
High truck traffic	X	X	X
Storm water runoff	X	X	X
COMMUNITY NEEDS			
Access to health care services	X	—	—
Access to healthy food	—	—	X
Access to the waterfront	X	X	—
Access to affordable housing	X	X	X
Access to public health facilities	—	—	X
Access to greenspace	X	X	X
Improved disaster preparedness and evacuation planning	X	X	X
Protection of local employment	—	X	X

plants, waste transfer stations, bus depots, and heavy truck traffic). In all three neighborhoods, many industrial facilities or former industrial sites are located on the waterfront, which makes them vulnerable to extreme flooding and heavy storm surges (Fainstein 2018; Bautista, Osorio, and Dwyer 2015). These neighborhoods and their residents are concerned about having adequate emergency preparedness capacity and evacuation centers during extreme weather events (NYCEJA, 2018). Low-income residents must bear the health consequences of living in proximity to these toxic sites. There is significant concern regarding toxic chemicals on the waterfront being displaced into residential areas (Madrigano et al. 2018; Bautista, Osorio, and Dwyer 2015). On the other hand, many young children and adults suffer from asthma and other respiratory illnesses, which can be exacerbated by worsened air quality during extreme heat events (Rosenthal, Kinney, and Metzger 2014). Due to a lack of quality recreational green and open space, the more vulnerable residents such as the elderly and children are at risk for heat-related illnesses (Rosenthal, Kinney, and Metzger 2014).

In order to address the unique ways and contexts in which communities are both ecologically and socially vulnerable, the CBOs on the CBA Workgroup emphasized that their communities lack some of the basic goods and services that are important to fostering resilient communities. They also emphasized how their lack of basic goods is connected to the legacy of systemic racism, exclusion, and disinvestment in their communities. This legacy includes the history of racial zoning, redlining, and urban renewal/slum clearance programs that have contributed to the racial stratification and structural disinvestment that persists in metropolitan areas like New York City. Understanding this legacy is part of the "contextual equity" analysis in our framework. It explains why these communities face a shortage of affordable and quality housing stock, lack of adequate health care and public health facilities, and lack of access to healthy food and green spaces. These disparities undermine residents' ability to face and adapt to the environmental and climate stressors present in their communities. Expanding their access to basic social and environmental goods, and addressing the legacy effects of systemic racism, should be a critical part of adaptation planning in socially vulnerable communities.

The CBOs also emphasized the importance of early and meaningful engagement with public officials in all phases of development planning in their communities, including adaptation planning and implementation. Each of the CBOs has engaged, often extensively, in adaptation planning in their communities and with their residents. To build community preparedness to climate-related emergencies, UPROSE in Sunset Park created the Be a Block Captain program designed to train residents to serve as "block captains" during extreme weather events. Local volunteers are trained to implement climate resilience strategies, including taking inventory of who lives on their block, serving as point persons for neighbors in case of emergency, and coordinating climate adaptation workshops. Similarly, THE POINT in Hunts Point, in partnership with several city agencies, established the Be a Buddy Program, which aims to connect local volunteers with the most at-risk residents and educate the community members about climate preparedness. WEACT, through multiple planning workshops with community members, put together the Northern Manhattan Climate Action (NMCA) Plan. The NMCA Plan contains policy recommendations and local actions organized around four themes: energy democracy, emergency preparedness, social hubs, and public participation. WEACT also launched Solar Uptown Now, a campaign that enables residents in Northern Manhattan to purchase solar panels as a group to bring down the cost of power for participants. While these community-led actions represent significant progress in addressing climate change impacts and mitigation in frontline communities, the challenge is how to align these efforts with the city's adaptation planning processes. As we will discuss in the next section, robust community engagement is a critical element of procedural equity in climate change adaptation.

Procedural Equity in Adaptation Planning

We explored procedural equity by gathering CBO perspectives on New York City's practices in recent and ongoing adaptation planning efforts. New York City explicitly recognizes the need for procedural equity in adaptation planning. Some typical ways in which the city engages with communities in adaptation planning are community meetings, inclusion of community representatives and organizations as part of advisory boards, and public forums and workshops (Foster et al. 2019). Yet even for those communities sought out for their input and engagement in city-led adaptation and resilience-building processes, there is a perception that existing city outreach efforts are conducted in good faith but ultimately may miss some of the ways these communities are uniquely vulnerable. In particular, the CBOs perceive that they are asked for their input and engagement often after critical policy and design choices have been made, sometimes leaving little room for the groups to meaningfully shape development to meet the needs of their communities.

The CBOs offered a number of examples of recent resilience-building initiatives and development decisions that prioritize market-oriented development and ignore the equity implications of these efforts. The Hunts Point Lifelines Resiliency Project (City of New York 2013), for example, involved a year-long community engagement process that identified flood risk and resilient energy as priority areas. This process was perceived by THE POINT CBO as very structured and rigid, with limited room for community inputs and creative ideas. While the project is making headway toward a more economically viable coastline, community members expressed concern that the city's concept of resilience was overly focused on coastal protection and renewable energy to the exclusion of social concerns such as gentrification and displacement. Similarly, in Sunset Park the CBO expressed heightened concern that development and resilience projects initiated or approved by the city could potentially lead to or accelerate displacement of local residents. Specifically, the CBO pointed to the mayor's plan for a Made in N.Y. Campus to bring back manufacturing to the waterfront (Santore 2017). Community members expressed concern about limited communication from the city about this initiative and lack of community engagement in a visioning process about development of the waterfront in ways that do not lead to nor accelerate displacement of residents (Santore 2017). The CBO expressed interest in linking the Made in N.Y. Campus to a community-led regenerative energy hub project. However, the CBO also expressed concern that the city's rezoning proposals to accommodate commercial development would limit possibilities for such a project.

Another consistent area of concern for each of these communities, but particularly in Northern Manhattan and Sunset Park, is that city-initiated adaptation and resilience projects may pave the way for new waterfront development projects and high-rise construction marketed toward higher-income white residents (Gould and Lewis 2018). Rising property values and rents associated with "climate gentrification" would mean outmigration of long-term residents and the weakening of social networks and social capital, both of which are necessary for creating resilient communities (Anguelovski et al. 2019). Each CBO expressed a strong desire for city officials and initiatives to actively support residents through cooperative practices that build up social capital and therefore preserve vulnerable neighborhoods through equitable development practices. As Schlosberg, Collins, and Niemeyer (2017) observe, in planning for climate adaptation, "local community groups . . . do not operate in a risk management or simple resilience framework" but rather "focus more on . . . basic needs and capabilities of every day." The CBO's suggested that adaptation and resilience planning might entail stronger focus on community development (e.g., building schools, affordable housing, safer streets, and greening space) to reduce the potential of displacing longtime residents and be more responsive to the social sustainability of these communities.

Our discussions also revealed that resources and capacity—both the city's and that of the CBOs—are significant variables for collaborative and equitable engagement. In particular, more established or relatively well-resourced (e.g., foundation-supported) CBOs are able to not only engage in their own adaptation planning processes but also, when given the chance, substantively and substantially contribute to adaptation plans and implementation. This could include helping the city design adaptation plans and projects that do not duplicate existing community-based efforts but rather leverage them. For instance, WE ACT has engaged in extensive climate action planning with deep community engagement and a collaborative process of identifying vulnerabilities and adaptation needs. Out of that process has emerged a focus on "critical infrastructure" required for emergency preparedness and resilience. Elements of this vision for Northern Manhattan echo the type of secure and equitable future envisioned in the first chapter of this book—including community microgrids, community centers, cooling centers, senior centers, access to grocery stores/food, and access to refrigeration for medication in an emergency. WE ACT is also focused on "energy democracy"—the shift from centralized, corporate fossil fuel–generated energy to energy generated and governed by communities and one that supports local economies, energy security, and the health and well-being of the people within those communities. Given this extensive planning and engagement process in place in Northern Manhattan, there is potential for the city to leverage these efforts to implement adaptation and resilience projects that account for both contextual and procedural equity.

While there is strong support for the city's efforts to ensure procedural equity, there is a strong feeling among the CBOs that there is room for more meaningful and empowering inclusion of vulnerable communities (NYCEJA 2016, 2018). In particular, each CBO expressed interest in a more fully collaborative, coproduction model of equitable adaptation and resilience planning in which city officials work side by side with CBOs (and other actors) at the outset to design and implement climate adaptation and resilience planning. Working side by side with communities at the outset to identify critical and intersecting climate, environmental, and social concerns and to codesign and co-implement adaptation projects was seen as key for reducing the potential of displacing longtime residents and promoting the social sustainability of local communities. Although such approaches are beginning to be implemented in adaptation planning, work on related issues such as community-based land use planning suggests that involvement of local partners at all phases of the design and implementation process is critical for the success and endurability of these efforts (Foster and Iaione 2015).

Conclusion

Equity is a central component of sustainable and just adaptation planning efforts in cities. This chapter described a case study of equity in adaptation planning in New York City. The study adopted an equity framework that incorporated three key dimensions of equity including distributive, contextual, and procedural equity. Distributive equity, which emphasizes disparities across social groups, neighborhoods, and communities in vulnerability, adaptive capacity, and the outcomes of adaptation actions, was incorporated through social vulnerability mapping analysis of spatial patterns of vulnerability in the city. Contextual equity, which considers how social, economic, and political factors and processes contribute to vulnerability and shape adaptive capacity, was addressed through case studies of socially vulnerable communities. Procedural equity, which emphasizes the extent and robustness of public and community participation in adaptation planning and decision-making, was explored through work with three CBOs who identify areas where city adaptation planning efforts can be more collaborative and inclusive.

The case study of New York City suggests several additional areas where our equity framework might be applied in other cities facing adaptation challenges. In particular, the framework may be useful for aiding in decisions about how adaptation projects are selected, including identification of where projects are needed and how they can be collaboratively tailored to meet the needs of local communities. The framework can also help cities reveal equity issues that may potentially arise as adaptation projects are implemented, including fuller examination of the potential unintended consequences of these projects. The framework may also be applicable for use in city- and region-wide adaptation planning efforts. In the face of climate change, many cities are beginning to consider implementation of large-scale flood barriers and other region-wide adaptation projects. All three forms of equity identified in this chapter can potentially be applied to local and regional efforts to plan for just adaptation to climate change.

NOTES

This chapter draws from a study that was conducted as part of the New York City Panel on Climate Change, Third Assessment Report. The full results of the study are presented in Foster et al. (2019). Funding support for this work was provided by the NOAA Regional Integrated Science Assessment (RISA) under grant number NA15OAR4310147 to Consortium for Climate Risk in the Urban Northeast (CCRUN).

REFERENCES

Abedi, V., Olulana, O., Avula, V., Chaudhary, D., Khan, A., Shima, S., et al (2021). Racial, economic, and health inequality and COVID-19 infection in the United States. *Journal of Racial and Ethnic Health Disparities*, 8(3), 732–742.

Adams, M. H. (2016). The end of Black Harlem. *New York Times*, May 27. https://www.nytimes.com/2016/05/29/opinion/sunday/the-end-of-black-harlem.html.

Adger, W. N. (2006). Vulnerability. *Global Environmental Change*, 16(3), 268–281.

Anguelovski, I., Connolly, J. J., Pearsall, H., Shorky, G., Checker, M., Maantay, J., et al. (2019). Opinion: Why green "climate gentrification" threatens poor and vulnerable populations. *Proceedings of the National Academy of Science*, 116(52), 26139–26143.

Austensen, M., Gould, E. I., Herrine, L., Karfunkel, B., Jush, G. T., and Moriarty, S. (2015). State of New York City's Housing and Neighborhoods. Furman Center for Real Estate and Urban Policy.

Bakkensen, L. A., Fox-Lent, C., Read, L. K., and Linkov, I. (2017). Validating resilience and vulnerability indices in the context of natural disasters. *Risk Analysis*, 37(5), 982–1004.

Bautista, E., Osorio, J. C., and Dwyer, N. (2015). Building climate justice and reducing industrial waterfront vulnerability. *Social Research: An International Quarterly*, 82(3), 821–838.

Bullard, R. D., and Alston, D. A. (1990). *We Speak for Ourselves: Social Justice, Race, and Environment*. Panos Institute.

Center for Disease Control Social Vulnerability Index (CDC SVI). (2018). SVI CDC/ATSDR SVI Publications and Materials. Agency for Toxic Substances and Disease Registry. https://www.atsdr.cdc.gov/placeandhealth/svi/publications /publications_materials.html.

Checker, M. (2011). Wiped out by the "Greenwave": Environmental gentrification and the paradoxical politics of urban sustainability. *City and Society*, 23(2), 210–229.

Chu, E., Anguelovski, I., and Carmin, J. (2016). Inclusive approaches to urban climate adaptation planning and implementation in the Global South. *Climate Policy*, 16(3), 372–392.

City of New York. (2013). PlaNYC: A Stronger, More Resilient New York. https://www1 .nyc.gov/site/sirr/report/report.page.

Cole, L. W., and Foster, S. R. (2001). *From the Ground Up: Environmental Racism and the Rise of the Environmental Justice Movement*. NYU Press.

Corburn, J., Osleeb, J., and Porter, M. (2006). Urban asthma and the neighborhood environment in New York City. *Health Place*, 12(2), 167–179.

Cornell, S., Berkhout, F., Tuinstra, W., Tàbara, D., Jäger, J., Chabay, I., et al. (2013). Opening up knowledge systems for better responses to global environmental change. *Environmental Science and Policy*, 28, 60–70.

Cutter, S. L., Boruff, B. J., and Shirley, W. L. (2003). Social vulnerability to environmental hazards. *Social Science Quarterly*, 84(2), 242–260.

Cutter, S. L., and Emrich, C. T. (2006). Moral hazard, social catastrophe: The changing face of vulnerability along the hurricane coasts. *Annals of the American Academy of Political and Social Sciences*, 604(1), 102–112.

Cutter, S. L., Emrich, C. T., Webb, J. J., and Morath, D. (2009). Social vulnerability to climate variability hazards: A review of the literature. *Final Report to Oxfam America*, 5, 1–44.

Cutter, S. L., and Finch, C. (2008). Temporal and spatial changes in social vulnerability to natural hazards. *Proceedings of the National Academy of Sciences*, 105(7), 2301–2306.

Cutter, S. L., Mitchell, J. T., and Scott, M. S. (2000). Revealing the vulnerability of people and places: A case study of Georgetown County, South Carolina. *Annals of the Association of American Geographers*, 90(4), 713–737.

Deas, M., Grannis, J., Hoverter, S., and DeWeese, J. (2017). Opportunities for Equitable Adaptation in Cities: A Workshop Summary Report. Georgetown Climate Center.

de Sherbinin, A. (2014). Climate change hotspots mapping: What have we learned?. *Climatic Change*, 123(1), 23–37.

Dunning, C. M., and Durden, S. E. (2011). Social Vulnerability Analysis Methods for Corps Planning. U.S. Army Engineer Institute for Water Resources.

Fainstein, S. S. (2018). Resilience and justice: Planning for New York City. *Urban Geography*, 39(8), 1268–1275.

Flanagan, B. E., Gregory, E. W., Hallisey, E. J., Heitgerd, J. L., and Lewis, B. (2011). A social vulnerability index for disaster management. *Journal of Homeland Security and Emergency Management*, 8(1), 11–22.

Foster, S. (1998). Justice from the ground up: Distributive inequities, grassroots resistance, and the transformative politics of the environmental justice movement. *California Law Review*, 86, 775.

———. (2017). Vulnerability, equality, and environmental justice: The potential and limits of law. In R. Holifield, J. Chakraborty, and G. Walker (Eds.), *The Routledge Handbook of Environmental Justice*. Routledge.

Foster, S. R., and Iaione, C. (2015). The city as a commons. *Yale Law and Policy Review*, 34, 281.

Foster, S., Leichenko, R., Nguyen, K. H., Blake, R., Kunreuther, H., Madajewicz, M., et al. (2019). New York City Panel on Climate Change 2019 Report, chapter 6: Community-Based Assessments of Adaptation and Equity. *Annals of the New York Academy of Science*, 1439(1), 126–173.

Fothergill, A., Maestas, E. G., and Darlington, J. D. (1999). Race, ethnicity, and disasters in the United States: A review of the literature. *Disasters*, 23(2), 156–173.

Fraser, N. (2009). *Scales of Justice: Reimagining Political Space in a Globalizing World*. Columbia University Press.

Gaynor, T. S., and Wilson, M. E. (2020). Social vulnerability and equity: The disproportionate impact of COVID-19. *Public Administration Review*, 80(5), 832–838.

Gonzalez, D. (2016). In Sunset Park, a call for "innovation" leads to fears of gentrification. *New York Times*, March 6. https://www.nytimes.com/2016/03/07/nyregion/in-sunset-park-a-call-for-innovation-leads-to-fears-of-gentrification.html.

Gould, K. A., and Lewis, T. L. (2018). From green gentrification to resilience gentrification: An example from Brooklyn. *City and Community*, 17(1), 12–15.

Hazard Vulnerability Research Institute (HVRI). (2018). Applications: Selected Applications of the Usage of SoVI. University of South Carolina. http://artsandsciences.sc.edu/geog/hvri/applications.

Kinney, P. L., Matte, T., Knowlton, K., Madrigano, J., Petkova, E., Weinberger, K., et al. (2015). New York City Panel on Climate Change 2015 Report Chapter 5: Public Health Impacts and Resiliency. *Annals of the New York Academy of Science*, 1336, 67–88.

Kirchhoff, C. J., Lemos, M. C., and Dessai, S. (2013). Actionable knowledge for environmental decision making: Broadening the usability of climate science. *Annual Review of Environment and Resources*, 38, 393–414.

Leichenko, R. (2011). Climate change and urban resilience. *Current Opinion in Environmental Sustainability*, 3(3), 164–168.

Leichenko, R., Klein, Y., Panero, M., Major, D. C., and Vancura, P. (2011). Equity and economics. *Annals of the New York Academy of Science*, 1244, 62–78.

Leichenko, R., McDermott, M., Bezborodko, E., Brady, M., and Namendorf, E. (2014). Economic vulnerability to climate change in coastal New Jersey: A stakeholder-based assessment. *Journal of Extreme Events*, 1(1), 1450003.

Lemos, M. C., and Morehouse, B. J. (2005). The co-production of science and policy in integrated climate assessments. *Global Environmental Change*, 15(1), 57–68.

Madrigano, J., Osorio, J. C. Bautista, E., Chavez, R., Chaisson, C. F., Meza, E., et al. (2018). Fugitive chemicals and environmental justice: A model for environmental monitoring following climate-related disasters. *Environmental Justice*, 11(3), 95–100.

McArdle, A. (2013). Storm surges, disaster planning, and vulnerable populations at the urban periphery: Imagining a resilient New York after superstorm Sandy. *Idaho Law Review*, 50, 19–46.

McDermott, M., Mahanty, S., and Schreckenberg, K. (2013). Examining equity: A multidimensional framework for assessing equity in payments for ecosystem services. *Environmental Science and Policy*, 33, 416–427.

Meerow S., Pajouhesh, P., and Miller, T. R. (2019). Social equity in urban resilience planning. *Local Environment*, 24(9), 793–808.

Myers, C. A., Slack, T., and Singelmann, J. (2008). Social vulnerability and migration in the wake of disaster: The case of Hurricanes Katrina and Rita. *Population and Environment*, 29(6), 271–291.

NAACP. (2015). Equity in Building Resilience in Adaptation Planning. https://naaee.org /sites/default/files/equity_in_resilience_building_climate_adaptation_indicators _final.pdf.

New York City Environmental Justice Alliance (NYCEJA). (2016). NYC Climate Justice Agenda: Strengthening the Mayor's OneNYC Plan. https://www.nyc-eja.org/wp -content/uploads/2017/10/CJA_041916.pdf.

———. (2018). NYC Climate Justice Agenda. Midway to 2030: Building Resiliency and Equity for a Just Transition. https://www.nyc-eja.org/wp-content/uploads/2018/04 /NYC-Climate-Justice-Agenda-Final-042018-1.pdf.

Preston, B. L., Yuen, E. J., and Westaway, R. M. (2011). Putting vulnerability to climate change on the map: A review of approaches, benefits, and risks. *Sustainability Science*, 6(2), 177–202.

Reckien, D. (2018). What is in an index? Construction method, data metric, and weighting scheme determine the outcome of composite social vulnerability indices in New York City. *Regional Environmental Change*, 18(5), 1439–1451.

Ribot, J. (2014). Cause and response: Vulnerability and climate in the Anthropocene. *Journal of Peasant Studies*, 41(5), 667–705.

Rosenthal, J. K., Kinney, P. L., and Metzger, K. B. (2014). Intra-urban vulnerability to heat-related mortality in New York City, 1997–2006. *Health and Place*, 30, 45–60.

Rosenzweig, C., Solecki, W. D., Blake, R., Bowman, M., Faris, C., Gornitz, V., et al. (2011). Developing coastal adaptation to climate change in the New York City infrastructure-shed: Process, approach, tools, and strategies. *Climatic Change*, 106(1), 93–127.

Sadd, J. L., Pastor, M., Morello-Frosch, R., Scoggins, J., and Jesdale, B. (2011). Playing it safe: Assessing cumulative impact and social vulnerability through an environmental justice screening method in the South Coast Air Basin, California. *International Journal of Environmental Research and Public Health*, 8(5), 1441–1459.

Sandy Regional Assembly. (2013). Recovery from the Ground Up: Strategies for Community-Based Resiliency in New York and New Jersey. http://newyork.resiliencesystem.org/sandy-regional-assembly-recovery-agenda.

Santore, J. (2017). Sunset Park officials, activists call for more engagement on development projects (updated). *Sunset Park Patch*, February 16. https://patch.com/new-york/sunset-park/sunset-park-officials-activists-call-more-engagement-development-projects.

Sarzynski, A. (2015). Public participation, civic capacity, and climate change adaptation in cities. *Urban Climate*, 14, 52–67.

Schlosberg, D. (2013). Theorising environmental justice: The expanding sphere of a discourse. *Environmental Politics*, 22(1), 37–55.

Schlosberg, D., and Collins, L. B. (2014). From environmental to climate justice: Climate change and the discourse of environmental justice. *Wiley Interdisciplinary Reviews: Climate Change*, 5(3), 359–374.

Schlosberg, D., Collins, L. B., and Niemeyer, S. (2017). Adaptation policy and community discourse: Risk, vulnerability, and just transformation. *Environmental Politics*, 26(3), 413–437.

Schmidtlein, M. C., Deutsch, R. C., Piegorsch, W. W., and Cutter, S. L. (2008). A sensitivity analysis of the social vulnerability index. *Risk Analysis: An International Journal*, 28(4), 1099–1114.

Sealey-Huggins, L. (2018). The climate crisis is a racist crisis: Structural racism, inequality, and climate change. In A. Johnson, R. Joseph-Salisbury, B. Kamunge, and G. Yancy (Eds.), *The Fire Now: Anti-Racist Scholarship in Times of Explicit Racial Violence* (p. 10). Bloomsbury.

Smit, B., and Wandel, J. (2006). Adaptation, adaptive capacity, and vulnerability. *Global Environmental Change*, 16(3), 282–292.

Solecki, W. (2012). Urban environmental challenges and climate change action in New York City. *Environment and Urbanization*, 24(2), 557–573.

Sze, J., and Yeampierre, E. (2018). Just transition and Just Green Enough: Climate justice, economic development and community resilience. In W. Curran and T. Hamilton (Eds.), *Just Green Enough: Urban Development and Environmental Gentrification* (pp. 61–73). Routledge.

Tate, E. (2012). Social vulnerability indices: A comparative assessment using uncertainty and sensitivity analysis. *Natural Hazards*, 63(2), 325–347.

U.S. Environmental Protection Agency EPA. (1992). Environmental Equity: Reducing Risks for All Communities. https://www.epa.gov/sites/default/files/2015-02/documents/reducing_risk_com_vol1.pdf.

CHAPTER 7

Making Movements

Mobilizing for More Just
Socioecological Futures in a Megacity

KIAN GOH

Environmental injustice in cities is now exacerbated by the impacts of climate change, as well as actions taken in response to climate change. Marginalized residents in cities, often Black, Indigenous, and people of color in the United States and poor, working-class people around the world, have frequently faced disproportionate environmental harms, such as proximity to toxic and polluting sources or lack of access to environmental amenities. Now these groups face not only the disparate impacts of climate change and environmental degradation but also the potentially unjust outcomes of plans and policies implemented in response to such impacts. Efforts to "green" or protect once-neglected sites can directly displace poor residents and cause gentrification, further excluding poor residents pushed to vulnerable sites by prior waves of urban development and exclusionary land policies (Gould and Lewis 2017; Anguelovski et al. 2019; Long and Rice 2019; Rice, Levenda, and Long, this volume).

Researchers have asserted that the most promising approaches to resisting both the impacts of climate change and unjust actions in response are those that are politically organized from "below"—from the point of view of systemically marginalized groups—and coordinated across a broader range of activists, sites, constituents, and collaborators (Routledge, Cumbers, and Derickson 2018; Agyeman et al. 2016; see also Goh 2020b, 2021). The viewpoint from below is particularly important when making claims of justice against structural oppression, while the coordination is essential if such struggles are to extend beyond specific places and threats to challenge broader systemic social, political, and environmental inequities. How do these movements from below do it?

This chapter explores the emerging practices of more just urban socioecological movement building in the context of climate change. In it, I focus on the multilevel and multiscalar networks and organizations necessary for climate-just praxis, in particular on how specific relationships among activists, community members, and researchers and professionals are mobilized. I develop the key concepts of this praxis based on my research in Jakarta, Indonesia, a city witness to severe environmental challenges and sociopolitical conflicts, where residents in informal kampong settlements face threats both from flooding and urban development. Here, social and environmental justice movements form networks to learn across different sites and social and environmental challenges, and develop strategies of action across multiple modes of knowledge and scales of activism. I explain the ways in which a broad coalition across places, actors, and knowledge modes is able to learn from and offer lessons for a more reflexive and expansive model of socioenvironmental movement building from the ground up.[1] I also explain my own reflexive approach as a researcher, forging sustained collaborative relationships on the ground while also following the extended networks and expanded terrain of urban climate struggles.

Green Exclusions and Urban Socioecological Movements

The increasing urgency of climate change has transformed our understanding of cities, the environment, and urban governance. In the face of increasingly evident and serious environmental impacts, and the growing attention to sustainable development, some cities have taken steps to mitigate the causes of and adapt to the impacts of climate change. A now deep and cross-disciplinary set of literatures attempts to explain the relationship between the factors of climate change and the form and function of cities, as well as how city plans and policies can mitigate the causes of climate change by reducing greenhouse gas emissions, or take steps to adapt to its impacts (see, e.g., Davoudi et al. 2009; Rosenzweig et al. 2011; Bulkeley 2013; Carmin, Dodman, and Chu 2013; Solecki et al. 2015).

As attention has turned toward cities' actions around climate change, questions and concerns about issues of justice have emerged (Romero-Lankao et al. 2018; Bulkeley 2021). This is the case whether on a global scale, where poorer nations of the Global South—those historically least responsible for carbon emissions—are suffering first and most from climate impacts, or within nation-states, where marginalized people—poor communities and communities of color in North America and Western Europe, and poor, working-class, Indigenous, and informal economy communities around the world—disparately bear the brunt from worsening health conditions and increasing climate-related disasters (Adger 2001; Roberts and Parks 2009). Such concerns have motivated an expansion of environmental justice concepts to address the factors, scopes, and scales of climate change (Sze and London 2008; Schlosberg and Collins 2014; Agyeman et al. 2016).

Increasingly, researchers have asserted how actions taken by cities in response to the unequal impacts of climate change can further exacerbate existing inequalities. This can occur whether by protecting resource flows or centers of economic productivity (Hodson and Marvin 2010), or, alternately, both taking actions and neglecting to take actions to manage ostensibly vulnerable areas and populations (Anguelovski et al. 2016). Urban sustainability plans and policies can lead to "green gentrification" or "climate gentrification," projects for climate mitigation, protection, or environmental benefits having the effect of displacing poor urban residents from once-vulnerable places (Checker 2011; Gould and Lewis 2017; Anguelovski et al. 2019; Rice et al. 2020). The reorientation of urban governance in response to climate change, with city governments and political-economic elites positioning cities as key sites of climate action, protecting privileged places and economic sectors, often perpetuates or worsens long-standing inequalities. Observations like these have led to an emerging discourse of "climate urbanism" (Long and Rice 2019; Castán Broto, Robin, and While 2020).

Such a context has made more critical the agenda around climate justice and cities—that is, the ways in which urban processes and socioenvironmental injustices are understood, and how more just approaches to urban climate change responses are conceived (in addition to this book, see Steele et al. 2012; Bulkeley et al. 2013; Goh 2020b). Given the ways in which urban responses to climate change have perpetuated uneven, exclusionary modes of urban development and consolidated political and economic power, the search for more just climate outcomes hinges on the possibility of alternative modes of urban development, different ways to conceive of climate urbanism. Researchers, for example, have posed the possibility, and necessity, of ground-up social movements organizing to challenge the injustices of development-as-usual responses to the climate emergency (Roberts and Parks 2009; Routledge, Cumbers, and Derickson 2018; Agyeman et al. 2016). But clear cases of—and conceptually rich explanations of—such movements on the ground have remained too rare.[2] This chapter seeks to address this by focusing on the intertwined processes of knowledge production and countermovement practices.

The chapter elaborates on the narratives of urban social movements in the context of such green exclusionary dynamics. It explains the ways in which marginalized actors and institutions on the ground mobilize to gain power across space and time, building broader grassroots-based networks to learn from different sites and environmental challenges, to develop ways of knowing across disciplines and ways of acting across different scales. It shows how reconfigured social relationships and institutional formations, organized from below, might lead to new, plural, or multiperspectival approaches to more just urban futures (cf. Nightingale et al. 2020).

Urban Environmental Problems in Jakarta

Jakarta, the capital of Indonesia, is a quintessential megacity. With a population of about 10 million in the capital proper and 30 million in the urban region, it faces compounding problems of fragmented governance, environmental degradation, "splintered" infrastructural provision, and social inequality. The capital region confronts chronic flooding problems. It experiences some level of flooding every year, with severe floods typically occurring every five to seven years. Within the last decades, such major inundations affected the capital district in 2002, 2007, 2013, and 2020, when, sometimes, a third of DKI Jakarta (Daerah Khusus Ibukota, the special capital region) was under water.

The problems of flooding in Jakarta are multivalent and complex. Most directly, they are due to the flooding from the rivers and canals during heavy rains and the flooding from the sea along the north coast of the city. But looking at the situation more broadly, the flooding is related to an expanded, and sometimes indirect, set of factors, including the geomorphological, climatic, infrastructural, and developmental conditions of the urban region. The Jakarta region is situated in a low-lying plain between the Jakarta Bay to the north and volcanic mountains to the south on the island of Java. Thirteen rivers flow through the capital city region northward from the mountains to the sea (figure 7.1). During heavy rains, rainwater falling along the length of the watersheds overcomes the rivers and canals, causing floods in the city. Global and regional climatological and meteorological forces bring annual periods of heavy rains, as well as cyclical tide patterns over decades, resulting in times of especially high tides that overflow the existing seawalls along the north coastline. Sea levels in the bay are also rising due to global warming, worsening the flooding from the sea.

Such geomorphological and climatic conditions are compounded by the infrastructural development of the urban region. Jakarta has attempted to control water with walls and canals since the early days of the Dutch colony Batavia in the early 1600s. Such hard infrastructural measures have continued through the centuries, with prolonged and often troubled projects to build

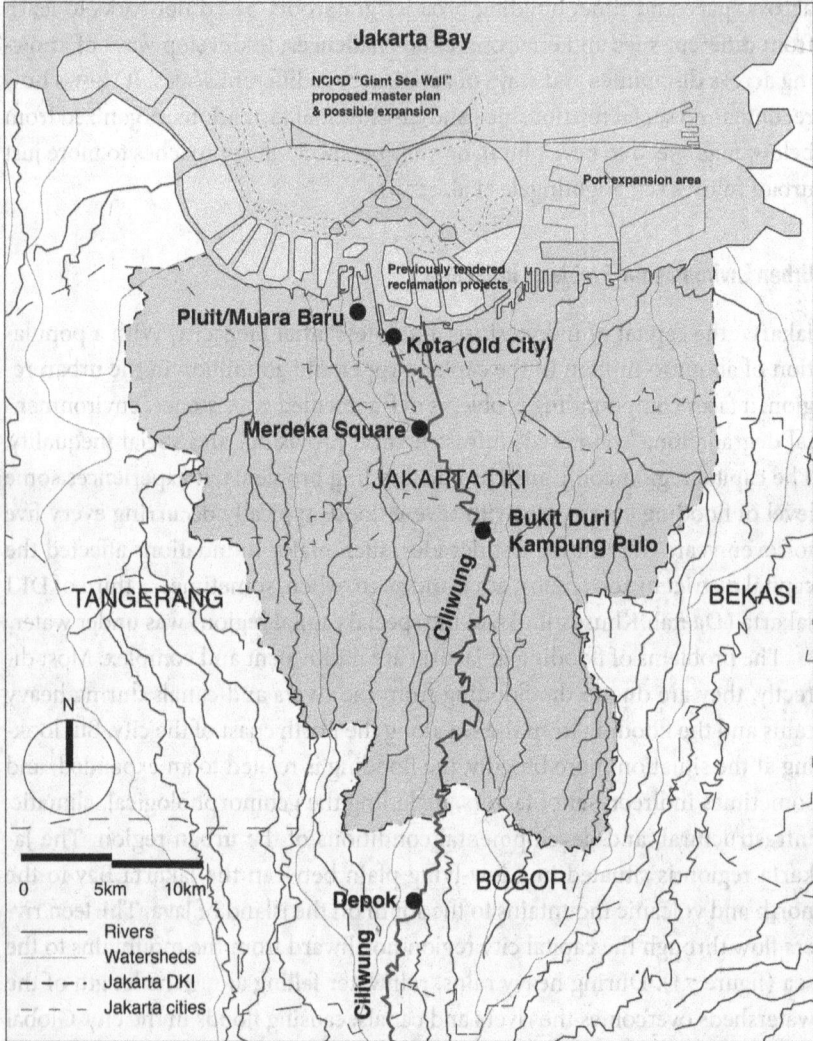

FIGURE 7.1. Jakarta showing major rivers, sites discussed in the chapter, and the NCICD master plan as originally proposed. Map by author

major flood canals (see Caljouw, Nas, and Pratiwo 2005; Simanjuntaka et al. 2012; Ward et al. 2013). The water management system, already choked by accumulations of trash and sediment, now faces a new challenge: rapid land subsidence. Parts of the city are sinking at rates of up to 12 centimeters a year along the coast (Brinkman and Hartman 2009; Abidin et al. 2011). This sinking—caused primarily by the overpumping of groundwater and exacerbated by rapid urban development and decreasing land permeability—is projected to lead to catastrophic flooding conditions by the mid-2030s. The flooding and subsidence has led, in recent years, to dredging and widening projects undertaken by the World Bank and the Japan International Cooperation Agency (JICA), and also to more ambitious proposals, including the National Capital Integrated Coastal Development (NCICD) master plan, known as the Giant Sea Wall. A Dutch-Indonesian initiative, the master plan called for a new city for 1.5 million people on reclaimed land spanning a portion of the Jakarta Bay, the outer wall forming outstretched wings in the shape of the Garuda, the mythical eagle that is the national symbol of Indonesia.

Further, the floods impact people and places in uneven ways. People living in the informal kampung settlements along the edges of the watercourses and the coastline are generally the most vulnerable. Many of these settlements, essentially urban villages, have been in these locations for decades, confined in place by racist colonial practices and further relegated to riskier areas and pressured by development in the nation-building and urban modernization decades following Indonesian independence (see Silver 2008; Putri 2019b). The ongoing dredging and widening projects directly threaten the settlements. Kampung residents are often seen as "illegal" and outside the norms of city life by city leaders and political and economic elites, even though they are part of urban social and economic life. City officials often blame kampung residents for the floods, saying their actions degrade the waterways, while ignoring rampant development and large-scale, illegal dumping.

Kampungs have long faced challenges from urban development and the heavy hand of government actions. But in recent years these conflicts around degrading environmental conditions and the projected impacts of climate change come alongside major political economic shifts in the city and region. While the end of the authoritarian Suharto regime in 1998 brought democratization and increasing political openness accompanied by persistent social differentiation and uneven development (see, e.g., Kusno 2013), the election of Joko Widodo (known as Jokowi) as governor of DKI Jakarta in 2012 seemed to herald a new era of state and society. The populist Jokowi espoused government reform and greater transparency and doubled down on liberalized urban development while embracing ideas about the ideal green and modern city. He also opened more fluid lines of communication between city officials and residents, including kampung communities, garnering, at least momentarily, a brightening relationship with kampung activists.

In this context, efforts to evict and demolish the informal settlements have been justified by the objective of more open green space, and to protect them against more destructive floods from the sea and the rivers and canals (Goh

2019b; see also Kusno 2013). Jokowi won his bid for the Indonesian presidency in 2014, after only two years as Jakarta's governor. His successor, Basuki Tjahaja Purnama (known as Ahok), largely continued Jokowi's urban development agenda, but often with a more contentious relationship with kampung residents. The dynamics of power, state-society relationships, and urban ideas, alongside the environmental changes in the city, have shifted the terrain of contestation, revealing new threats against the settlements as well as new arenas for kampung activism.

Socioecological Movements in Jakarta

I arrived in Jakarta in January 2013, after another catastrophic flood, eager to research the ways in which urban planners, designers, engineers, and grassroots activists were taking on the serious social and environmental challenges they faced. I was immediately advised by an architect friend from the city to contact the Rujak Center for Urban Studies. Through Rujak, I was introduced to organizers from the Urban Poor Consortium (UPC) and Ciliwung Merdeka, two groups working with kampung activists, who invited me to witness their community organizing meetings. The research I began that year eventually became part of a larger and longer-term project on the spatial politics of urban climate change responses, with fieldwork sites in New York and Rotterdam in addition to Jakarta, encompassing on-the-ground participant observation, approximately fifty-five semistructured interviews across the sites, and reviews of about thirty-five planning and design documents, primarily conducted between 2013 and 2019.

Through this ongoing and expanded research, the grounded and positional nature of my engagements has remained. My continued researcher–key informant relationship with individuals from Rujak, UPC, and Ciliwung Merdeka is cultivated in ways that are partially identity based (as a Bahasa Indonesia-speaking person originally from Southeast Asia), partially political (my support of the transformative potential of social movements affirmed through everyday interactions and the occasional presentations of my work during lectures hosted by Rujak), and partially of privilege (my claims to legitimacy bolstered by my position, back then as a PhD student at the Massachusetts Institute of Technology and now as a faculty member at the University of California, Los Angeles). This latter aspect surely also gained me access to high-level decision makers in the city, enabling me to conduct research interviews with officials from DKI and national governments as well as professional consultants based in Jakarta and in the Netherlands. While my privilege and partial familiarity also helped access with organizers, it nevertheless maintained clear class differences with kampung residents, with whom I interacted primarily as a researcher versed in place and culture, but from somewhere else. This multivalent frame offered by positionality and privilege afforded my view of close, on-the-ground dynamics as well as the workings of global networks of expertise, critical to contextualizing and understanding even very local conflicts (see Goh 2019a, 2020a).

In Jakarta the responses to the multivalent threats among social and environmental justice movements have been exemplary. In a number of kampung areas, community activists have been organizing against eviction and displacement. They have formed innovative coalitions, produced alternative urban design ideas, launched legal actions, and negotiated agreements with political campaigns in order to shift the narrative of the contestation, and to gain power and make claims on space.[3] Here, I illustrate four notable examples.

URBAN POOR CONSORTIUM'S COALITION BUILDING IN MUARA BARU

In North Jakarta, around the Pluit Reservoir, a large catchment pond that holds water from rivers and canals before it is pumped into the Jakarta Bay, grassroots kampung activists from Kampung Muara Baru, organized by UPC, have been fighting against ongoing efforts to evict and demolish informal settlements as well as for rehousing in place for those who have been evicted. In 2013, as part of a city government project ostensibly for flood mitigation and more open, green space, a large swath of the settlement along the retention pond was demolished to make way for a new park. The park, called Taman Kota Waduk Pluit (Pluit Lake Park), is popularly known as Taman Jokowi (Jokowi's Park), a clear nod to the then-governor's role in greening and clearing the area.

UPC and grassroots activists from Jaringan Rakyat Miskin Kota (JRMK) (Urban Poor Network) worked with kampung residents to protest further demolitions, and, along with community architects Arkomjogja and students from University Indonesia, proposed a new model of more socially attuned housing, with spaces for more collective uses and informal economic activities. According to the late Edi Saidi, then UPC coordinator, their coalitions' ground-up organizing and a more open, participatory city governing style brought increased trust and dialogue between the city government and kampung communities. While continuing to criticize the efforts to demolish the settlements, he expressed guarded optimism about future dialogue and actions. According to Saidi, "There [in Muara Baru], we have process for dialogue, participation, as well as a longer-term view; a model of housing that meets their desires and needs."[4] However, as of 2018, while multiple new *rusunawa*, or social housing blocks, have been constructed in the area, no examples of the social housing models designed with kampung resident participation have been built.

CILIWUNG MERDEKA'S COUNTERPLANNING IN BUKIT DURI AND KAMPUNG PULO

In Central Jakarta, around two tight bends of the notorious Ciliwung River—the longest river, often associated with the worst flooding—activist groups working with residents in Kampung Bukit Duri and Kampung Pulo have protested projects to widen the river, concretize the riverbanks—to "normalize" them, in the words of city and development aid officials—and clear a riverfront maintenance access road. Such projects would entail the demolition of hundreds of housing structures along the river in the two kampungs. Here, the organization Ciliwung Merdeka (literally translated as "Free Ciliwung") worked with Kota Kita (a citizen planning group), planning and architecture student interns, and a community development intern from the Asian Coalition for Housing Rights (ACHR) to develop alternative approaches to minimize the destruction of the settlement. The coalition conducted a community mapping project, organizing community volunteers across the kampungs to gather information on issues such as ownership, land use, infrastructure, community spaces, and socially and culturally significant landmarks. The group also mapped the possible impact of the city officials' proposed river-widening regulations. Working with community architects, the coalition proposed an alternative river-widening scheme, envisioning a new, stacked kampung housing model affording a more dense settlement set away from the river's edge (see Shepherd 2013).

According to Sandyawan Sumardi, activist and leader of Ciliwung Merdeka, the coalition's work prioritized the "extraordinary social model" of the kampungs, their conjoining of live and work worlds, and their often invisible socioeconomic links across the city, far beyond their physical territory. He noted in particular that the documented plans and design proposals repositioned kampung residents' struggles. The collective planning process, and the concrete proposals, proved that marginalized communities could organize and assert agency and imagination about the future of their own people and places,[5] a "counterplan" against business-as-usual urban development. They garnered agreements with city officials to consider their alternative proposals. These efforts, however, did not prevent large parts of both kampungs from being forcibly evicted and demolished in 2015 (see Van Voorst and Padawangi 2015).

FIGURE 7.2. Ciliwung River with new concrete embankments and partially demolished buildings, May 2017. Photograph by author

In 2018 partially destroyed houses could still be seen on both sides of the Ciliwung, with the river now contained by raised concrete embankments and wide access roads (figure 7.2).

More recently, in Kampung Tongkol, not far from the Pluit Reservoir and close to Kota, the old center of the city, residents faced the same threats of eviction and demolition because of river-widening projects. Here, across 2015 and 2016, kampung residents and activists, along with organizers from UPC, organized themselves to conduct a voluntary, selective dismantling and reconfiguration of the settlement, physically reorienting the houses toward the river and allowing for maintenance access. According to UPC community organizer Gugun Muhammad, this involved deliberations with kampung residents about resizing and redistributing space, and collective agreements to standardize accommodations around minimum requirements.[6] These actions protected key stretches of the kampung from wholesale demolition.

The efforts here transcended physical change, striving for a social and spatial model of kampung empowerment. Kampung residents also started a waste management program and crafting and local agriculture projects. On one side of the river, they built a new model house, based on more sustainable materi-

FIGURE 7.3. Model house in Kampung Tongkol, May 2017. Photograph by author

als and construction, as a prototype for sustainable kampung redevelopment (figure 7.3). Architect Kamil Muhammad from the group Architecture Sans Frontières Indonesia (ASF-ID) explained that he had been brought on because of an existing working relationship with UPC's Gugun Muhammad (no relation) on a research initiative on participation. The architect explained how he, working with students from the University of Indonesia and with funding through ACHR, designed the model house as part of the broader social and spatial pattern of the reconfigured settlement, an example for what could be.[7] The strategy at Kampung Tongkol is largely considered among activists and kampung residents to be a successful example, showcasing residents' agency and proactivity, although it entailed that parts of the existing settlement were still demolished.

ORGANIZING FOR BROADER POLITICAL CHANGE

Finally, the coalition of kampung activists and organizers also set their sights on more sustained, urban, region-wide change. In the lead-up to the 2017 Jakarta gubernatorial election, pro-poor activist groups, coordinated by UPC, JRMK, and Rujak Center for Urban Studies, broached a political agreement with the campaign of challenger Anies Baswedan. Anies's campaign held policies counter to those implemented by Jokowi and his successor, Ahok. These policy positions were further clarified and solidified in the campaign's negotiations with kampung activists, including forty-six specific points about land titling and zoning, such as legalization of settlements in areas undergoing or under threat of evictions, as well as support for informal economic activities, such as financial assistance for street vendors and *becak* (rickshaw) drivers (see Savirani and Aspinall 2017). This was not the first time the activists had secured a political agreement; they had done so with Jokowi during the 2012 election as well. This time, they tried to rectify the shortcomings of the previous agreement, procuring the pact in writing and including more specific goals and concrete demands in it.

The Anies campaign ultimately prevailed in the election, and in the following year the new governor initiated a series of programs meant to improve conditions in kampungs (Colven and Irawaty 2019). Notably, the gubernatorial decree number 878 of 2018 issued by Anies declared that participatory community action plans would be undertaken in twenty-one kampungs (including Kampung Tongkol, previously mentioned). While not necessarily an unmitigated success—with kampung organizers still challenging the new governor's commitment in the months after—the actions around the electoral victory show how kampung organizers have made continual adjustments to their strategies of making concrete demands and building coalitions.

Praxis in the Megacity

In the face of continual challenges, pro-poor coalitions in Jakarta have fought back, garnered some political power, harnessed institutional and legal frameworks to push forward agency, and developed concrete proposals—alternative ideas about the social and spatial future of the city. Their praxis encompasses what might be called a multivalency: building movements on the ground, developing collaborations, harnessing expertise, and negotiating modes of engagement. How have these movements in Jakarta done this? Two factors are especially important. Together, they enable the movements to take on the historically and spatially constituted socioecological conditions they face.

First, this movement is part of broader networks, learning from diverse sites and social and environmental challenges over time. UPC established the Urban Poor Linkage (UPLINK), a network across fourteen Indonesian cities, in 2002. This network has functioned as a political alliance as well as an exchange of expertise and knowledge, broadening a base of pro-poor social movements across key Indonesian cities. Two examples of this are particularly pertinent. In 2004 the Indian Ocean tsunami, caused by an earthquake off the northwestern coast of Sumatra island in Indonesia, devastated the Aceh Province, including Banda Aceh, the major city in the region. Uplink Banda Aceh (UBA) formed soon after the tsunami, conducted emergency response work in the immediate aftermath of the disaster, and provided community-based alternatives to the work of the larger, transnational nongovernmental aid groups during the reconstruction period. UBA helped form a coalition of twenty-three kampungs in the region, gathered data on demographics and land, and helped the villages undertake a community-based, consensus-driven replotting for reconstruction. The UBA coalition planned new models of more storm-resilient housing structures and methods of construction (see Syukrizal, Hafidz, and Sauter 2009; Vale, Shamsuddin, and Goh 2014).

Around the same period, in Surabaya, the second largest city on Java island, Uplink Surabaya was working with community organization Paguyuban Warga Strenkali Surabaya (PWS) (Riverside Community Association) in their efforts to stop the eviction and demolition of riverfront kampung settlements—local circumstances quite similar to those faced by kampung communities along the Ciliwung River in Jakarta. Uplink Surabaya conducted technical studies of river flows to challenge the government's own reports; with PWS, developed a participatory engagement effort with riverfront kampung residents; and, having built support from within and outside the affected communities, negotiated an alternative agreement with the city government. The communities agreed to proactively downsize their housing structures and reorient them to face the river, facilitating the widening and dredging of the waterway and allowing the construction of a waterfront access road—a strategy that would be echoed in the organizing and rebuilding in Kampung Tongkol in Jakarta. They also agreed to develop a community waste management system. The networked nature of Uplink's approach was evident, with organizing assistance and technical advice coming from groups in other cities—including architects who had been involved in the Banda Aceh posttsunami reconstruction work (see Some, Hafidz, and Sauter 2009).

In both Banda Aceh and Surabaya, community organizers from UPC, researchers and advocates who would go on to establish Rujak in Jakarta, and architects versed in community-engaged design worked alongside grassroots, place-based organizations. In both cases, the physical reconstruction work was not an end in itself but part of a broader effort at building solidarity and pushing for social change. This networked, knowledge-building approach enables the coalition of organizers to link challenges on multiple levels: on the ground, in response to direct threats, and across sets of parallel, overlapping, and distinct site and political conditions in various cities (see Goh 2021). In parallel to the ways in which large-scale, state-led urban environmental plans are increasingly constituted through global networks (see Goh 2020a), this networked approach enables the coalition to transcend some of the challenges typically faced by very local grassroots activists, enabling a broader sense of learning and sharing.

Second, the movement builds on multiple modes of knowledge and fronts of activism. As previously noted, the factors around flooding in Jakarta are complex and far reaching, involving geomorphological, biophysical, historical, and sociopolitical factors. Challenging convenient or straightforward narratives about the flooding that unjustly impact kampung residents requires more than specific, very local community-based activist responses. The networked nature of the movement, as already delineated, helps. But it also requires concerted attention to the various ways in which struggles might be constituted, including a better understanding of the different strategies of activism and approaches to knowledge production of practices from diverse sites.

Here the research and advocacy of an organization like Rujak, which bills itself as a center for both thinking and action, is critical. On one level, Rujak is a convener of individuals and groups invested in urban issues, particularly around social justice, housing rights, and environmental sustainability, and a research center and information clearinghouse on those topics. In 2017 the organization cohosted an event with Leilani Farha, then UN special rapporteur on adequate housing, centering the struggles of the urban poor in the city. It regularly holds events and publishes pamphlets and books on sustainable urban practices. I have participated as a guest speaker at two of these events. In the wake of the recent floods, Rujak disseminated government documents through its website, enabling activists and researchers to find often obscure presentations on the city's dredging and reclamation plans. On another level, it functions as a political advocacy organization, coordinating with community organizers at UPC and grassroots kampung activists to, for example, garner agreements with political parties and city officials.

Kampung activists have also maintained strong and ongoing relationships with politically allied professional and academic groups. I include myself as part of this group. In Bukit Duri and Kampung Pulo, activists launched legal challenges to the evictions, working with lawyers from Lembaga Bantuan Hukum Jakarta (Jakarta Legal Aid Society) (Shatkin and Soemarwi 2021). In each of the sites discussed earlier, coalitions included architects and designers, such as Arkomjogja and ASF-ID—architects who have expressly decided to take on and practice community-engaged and participatory methods.[8] The embrace of multiple knowledge modes extends to the ways in which members of the broader movement engage with researchers from Indonesia and beyond, helping make legible the political and environmental conflicts in the city from their point of view. Indeed, I was implored to reach out to Marco Kusumawijaya, then director of Rujak (and who had also been involved in the reconstruction work in Banda Aceh) while beginning my research in Jakarta. Through my relationship with Kusumawijaya, Dian Tri Irawaty (formerly of Rujak, now a PhD candidate studying urban politics in Jakarta), and Elisa Sutanudjaja, currently Rujak's executive director, I have sustained a close engagement with efforts on the ground even through various institutional leadership changes and, on my end, an increasingly multisited research initiative.

This multi–knowledge mode approach often works at the scale of the urban region, taking into consideration the ways in which socioecological challenges in the context of climate change do not stop at territorial borders or around particular local political conflicts. Part of Rujak's work, that of conducting historical and legal research that links land reclamation laws in the bay and dredging projects along the rivers and canals, makes visible the sociopolitical, biophysical-environmental, and infrastructural interconnections across space, and clarifies how such socioecological challenges are constituted across the urban ecological region, that is, the infrastructural and ecological watershed (see Goh 2019a). This recognition suggests that activists might build collaborative movements across the urban ecological region, with fights for development controls in the upper reaches of the watershed—where rain that is not infiltrated into the ground because of decreasing permeability is flushed into the rivers, worsening floods downstream—possibly aiding the fights against displacement in the central Jakarta kampungs.

Conclusion

In Jakarta, grassroots and community-based countermovements have succeeded in building power and opposing, albeit still in somewhat limited ways, unjust and exclusionary urban environmental and development actions driven by political-economic elites in the city. They have done this through innovative coalitions and organizing strategies, building networks of political alliance and organizing and technical expertise across space and time, and engagement across multiple knowledge modes and fronts of activism. Such networked, multi–knowledge mode, and urban regional approaches have enabled selectively successful challenges to power structures and protection of some local communities. Understanding the effectiveness of such coalitions is particularly important in light of the multiscalar nature of both climate change impacts and the possibilities of the emerging climate justice movement (Goh 2020b).

While the successes are fleeting at times, and the longer-term struggles are still ongoing, such movement building offers lessons for climate-just praxis. It helps conceptualize urban climate politics for sites, such as Jakarta, that confront often extreme social and environmental challenges and disparate spatial conditions. And as I tell it here, it underscores too the always-to-be-interrogated position of justice-oriented urban researchers within the constellation of urban socioecological movements.

NOTES

Thank you to Dian Tri Irawaty for continual guidance around the nuances of urban social movements in Jakarta.

1. The research in this chapter is a part of the author's larger project on the spatial politics of urban climate change responses in Jakarta, New York, and Rotterdam. This mixed-methods research involved field visits, participant observation, approximately fifty-five semistructured interviews, and review of thirty-five planning and design documents, conducted between 2013 and 2019 (see Goh 2021).

2. See, for example, Dodman, Mitlin, and Rayos (2010), and examples discussed later in this chapter.

3. A growing set of literature investigates and affirms the role of nongovernmental organizations and grassroots activists in Jakarta around the issue of urban flooding. I consider this chapter to be adding to and in conversation with them. See Padawangi and Douglass (2015), Padawangi et al. (2016), Savirani and Aspinall (2017), Batubara, Kooy, and Zwarteveen (2018), Leitner and Sheppard (2018), Colven and Irawaty (2019), Goh (2019a, 2019b), Putri (2019a), and Shatkin and Soemarwi (2020).

4. Interview by the author (translated from Bahasa Indonesia), Jakarta, July 15, 2014.

5. Interview by the author (translated from Bahasa Indonesia, Jakarta, July 18, 2014.

6. Interview by the author, Jakarta, December 11, 2018.

7. Interview by the author, Jakarta, December 17, 2018.

8. Interview by the author, Jakarta, December 17, 2018.

REFERENCES

Abidin, H. Z., Andreas, H., Gumilar, I., Fukuda, Y., Pohan, Y. E., and Deguchi, T. (2011). Land subsidence of Jakarta (Indonesia) and its relation with urban development. *Natural Hazards*, 59(3), 1753–1771.

Adger, N. W. (2001). Scales of governance and environmental justice for adaptation and mitigation of climate change. *Journal of International Development*, 13(7), 921–931.

Agyeman, J., Schlosberg, D., Craven, L., and Matthews, C. (2016). Trends and directions in environmental justice: From inequity to everyday life, community, and just sustainabilities. *Annual Review of Environment and Resources*, 41, 321–340.

Anguelovski, I., Connolly, J. T. J., Pearsall, H., Shokry, G., Checker, M., Maantay, J., et al. (2019). Opinion: Why green "climate gentrification" threatens poor and vulnerable populations. *Proceedings of the National Academy of Sciences*, 116(52), 26139–26143.

Anguelovski, I., Shi, L., Chu, E., Gallagher, D., Goh, K., Lamb, Z., et al. (2016). Equity impacts of urban land use planning for climate adaptation: Critical perspectives from the Global North and South. *Journal of Planning Education and Research*, 36(3), 333–348.

Batubara, B., Kooy, M., and Zwarteveen, M. (2018). Uneven urbanisation: Connecting flows of water to flows of labour and capital through Jakarta's flood infrastructure. *Antipode*, 50(5), 1186–1205.

Brinkman, J. J., and Hartman, M. (2009). Jakarta Flood Hazard Mapping Framework. http://edepot.wur.nl/140833.

Bulkeley, H. (2013). *Cities and Climate Change*. Routledge.

———. (2021). Climate changed urban futures: Environmental politics in the anthropocene city. Environmental Politics, 30, 1–19.

Bulkeley, H., Carmin, J., Castán Broto, V., Edwards, G. A. S., and Fuller, S. (2013). Climate justice and global cities: Mapping the emerging discourses. *Global Environmental Change*, 23(5), 914–925.

Caljouw, M., Nas, P. J. M., and Pratiwo, M. R. (2005). Flooding in Jakarta: Towards a blue city with improved water management. *Bijdragen tot de taal-, land- en volkenkunde*, 161(4), 454–484.

Carmin, J., Dodman, D., and Chu, E. (2013). *Urban Climate Adaptation and Leadership*. OECD Regional Development Working Papers. OECD Publishing.

Castán Broto, V., Robin, E., and While, A. (Eds.). 2020. *Climate Urbanism: Towards a Critical Research Agenda*. Palgrave Macmillan.

Checker, M. (2011). Wiped out by the "Greenwave": Environmental gentrification and the paradoxical politics of urban sustainability. *City and Society*, 23(2), 210–229.

Colven, E., and Irawaty, D. T. (2019). Critical spatial practice and urban poor politics: (Re)Imagining housing in a flood-prone Jakarta. *Society and Space*, http://societyandspace.org/2019/08/26/critical-spatial-practice-and-urban-poor-politics-reimagining-housing-in-a-flood-prone-jakarta/.

Davoudi, S., Crawford, J., and Mehmood, A. (Eds.). (2009). Planning for Climate Change: Strategies for Mitigation and Adaptation for Spatial Planners. Earthscan.

Dodman, D., Mitlin, D., and Rayos, J. (2010). Victims to victors, disasters to opportunities: Community-driven responses to climate change in the Philippines. *International Development Planning Review*, 32(1), 1–26.

Goh, K. (2019a). Urban waterscapes: The hydro-politics of flooding in a sinking city. *International Journal of Urban and Regional Research*, 43(2), 250–272.

———. (2019b). Toward transformative urban spatial change: Views from Jakarta. In T. Banerjee and A. Loukaitou-Sideris (Eds.), *The New Companion to Urban Design* (pp. 519–532). Routledge.

———. (2020a). Flows in formation: The global-urban networks of climate change adaptation. *Urban Studies*, 57(11), 2222–2240.

———. (2020b). Urbanising climate justice: Constructing scales and politicising difference. *Cambridge Journal of Regions, Economy and Society*, 13(3), 559–574.

———. (2021). *Form and Flow: The Spatial Politics of Urban Resilience and Climate Justice*. MIT Press.

Gould, K. A., and Lewis, T. L. (2017). *Green Gentrification: Urban Sustainability and the Struggle for Environmental Justice*. Routledge.

Hodson, M., and Marvin, S. (2010). World Cities and Climate Change: Producing Urban Ecological Security. Open University Press.

Kusno, A. (2013). *After the New Order: Space, Politics, and Jakarta*. University of Hawai'i Press.

Leitner, H., and Sheppard, E. (2018). From Kampungs to condos? Contested accumulations through displacement in Jakarta. *Environment and Planning A: Economy and Space*, 50(2), 437–456.

Long, J., and Rice, J. L. (2019). From sustainable urbanism to climate urbanism. *Urban Studies*, 56(5), 992–1008.

Nightingale, A. J., Eriksen, S., Taylor, M., Forsyth, T., Pelling, M., Newsham, A., et al. (2020). Beyond technical fixes: Climate solutions and the great derangement. *Climate and Development*, 12(4), 343–352.

Padawangi, R., and Douglass, M. (2015). Water, water everywhere: Toward participatory solutions to chronic urban flooding in Jakarta. *Pacific Affairs*, 88(3), 517–550.

Padawangi, R., Herlily, E. T., Prescott, M. F., Lee, I., and Shepherd, A. (2016). Mapping an alternative community river: The case of the Ciliwung. *Sustainable Cities and Society*, 20, 147–157.

Putri, P. W. (2019a). Insurgent planner: Transgressing the technocratic state of postcolonial Jakarta. *Urban Studies*, 57(9), 1845–1865.

———. (2019b). Sanitizing Jakarta: Decolonizing planning and Kampung imaginary. *Planning Perspectives*, 34(5), 805–825.

Rice, J. L., Cohen, D. A., Long, J., and Jurjevich, J. R. (2020). Contradictions of the climate-friendly city: New perspectives on eco-gentrification and housing justice. *International Journal of Urban and Regional Research*, 44(1), 145–165.

Roberts, J. T., and Parks, B. C. (2009). Ecologically unequal exchange, ecological debt, and climate justice: The history and implications of three related ideas for a new social movement. *International Journal of Comparative Sociology*, 50(3–4), 385–409.

Romero-Lankao, P., Bulkeley, H., Pelling, M., Burch, S., Gordon, D. J., Gupta, J., et al. (2018). Urban transformative potential in a changing climate. *Nature Climate Change*, 8(9), 754.

Rosenzweig, C., Solecki, W. D., Hammer, S. A., and Mehrotra, S. (Eds.). (2011). *Climate Change and Cities: First Assessment Report of the Urban Climate Change Research Network*. Cambridge University Press.

Routledge, P., Cumbers, A., and Derickson, K. D. (2018). States of just transition: Realising climate justice through and against the state. *Geoforum*, 88 (January), 78–86.

Savirani, A., and Aspinall, E. (2017). Adversarial linkages: The urban poor and electoral politics in Jakarta. *Journal of Current Southeast Asian Affairs*, 36(3), 3–34.

Schlosberg, D., and Collins, L. B. (2014). From environmental to climate justice: Climate change and the discourse of environmental justice. *Wiley Interdisciplinary Reviews: Climate Change*, 5(3), 359–374.

Shatkin, G., and Soemarwi, A. (2021). Risk and the dialectic of state informality: Property rights in flood-prone Jakarta. *Annals of the American Association of Geographers*, 111(4), 1183–1199.

Shepherd, A. (2013). Jakarta, Indonesia. In J. Brugman, B. Dovarch, Z. Kassam, F. Pasta, and A. Shepherd (Eds.), *Grounding Knowledge: Reflections on Community-Driven Urban Practices in South-East Asia* (pp. 50–81). Development Planning Unit (DPU), Community Architects Network (CAN), Asian Coalition for Housing Rights (ACHR). https://issuu.com/dpuucl/docs/grounding_knowledge_24.2.14_new.

Silver, C. (2008). *Planning the Megacity: Jakarta in the Twentieth Century*. Routledge.

Simanjuntaka, I., Frantzeskakib, N., Enserinka, B., and Ravesteijnc, W. (2012). Evaluating Jakarta's flood defence governance: The impact of political and institutional reforms. *Water Policy*, 14(4), 561–580.

Solecki, W., Seto, K. C., Balk, D., Bigio, A., Boone, C. G., Felix, C., et al. (2015). A conceptual framework for an urban areas typology to integrate climate change mitigation and adaptation. *Urban Climate*, 14 (December), 116–137.

Some, W., Hafidz, W., and Sauter, G. (2009). Renovation not relocation: The work of Paguyuban Warga Strenkali (PWS) in Indonesia. *Environment and Urbanization*, 21(2), 463–475.

Steele, W., MacCallum, D., Byrne, J., and Houston, D. (2012). Planning the climate-just city. *International Planning Studies*, 17(1), 67–83.

Syukrizal, A., Hafidz, W., and Sauter, G. (2009). Reconstructing Life after the Tsunami: The Work of Uplink Banda Aceh in Indonesia. International Institute for Environment and Development London.

Sze, J., and London, J. K. (2008). Environmental justice at the crossroads. *Sociology Compass*, 2(4), 1331–1354.

Vale, L. J., Shamsuddin, S., and Goh, K. (2014). Tsunami + 10: Housing Banda Aceh after disaster. *Places Journal*, December. https://placesjournal.org/article/tsunami-housing-banda-aceh-after-disaster/.

Van Voorst, R., and Padawangi, R. (2015). Floods and forced evictions in Jakarta. *New Mandala*, August 21. http://asiapacific.anu.edu.au/newmandala/2015/08/21/floods-and-forced-evictions-in-jakarta/.

Ward, P. J., Pauw, W. P., Van Buuren, M. W., and Marfai, M. A. (2013). Governance of flood risk management in a time of climate change: The cases of Jakarta and Rotterdam. *Environmental Politics*, 22(3), 518–536.

CHAPTER 8

Visibilizing Queer Resilience

Representational Justice for the Climate Movement

VANESSA RADITZ

The People's Climate March in September 2014 brought over three hundred thousand people to New York City (NYC) in solidarity with a global day of mobilization for climate action. In recognition of and respect for the decades of work by the environmental justice (EJ) movement, the NYC march was led by hundreds of environmental justice organizations from around the world. As they marched through downtown Manhattan, activists hoisted a banner that stretched across three lanes of the reclaimed street: "FRONTLINES OF CRISIS—FOREFRONTS OF CHANGE." This slogan speaks to the EJ movement's important work to uplift representation of the most vulnerable as the necessary leaders for the climate movement.

In the days leading up to the march, the Audre Lorde Project, a community organizing center in NYC dedicated to queer and trans Black and Indigenous people of color (QTBIPOC),[1] released a letter of solidarity with the climate movement. The letter highlighted climate vulnerability among QTBIPOC, especially the disproportionate number who are homeless or chronically ill, and they expressed their shared interest in uprooting the colonial and capitalist systems at the root of multiple crises. Our communities, they write, "live on the front lines" (Audre Lorde Project 2014). Yet when the lineup for the march was announced, they were dismayed to see LGBTQ+ people placed in the seventh and final group. As one of the coauthors of the Audre Lorde Project's letter wrote: "It was even behind groups who 'can build the future,' 'have solutions,' and those who 'know who is responsible.' This once again erases the narratives and legacies we carry . . . and re-opens the familiar wounds of marginalization from society and mainstream movements" (Pineda 2015). This erasure is particularly stark given the vast contributions of queer and trans people throughout the climate movement. Though they may not be represented in name among the front lines and forefronts, QTBIPOC even designed the banner bearing that slogan.[2]

The climate disaster experiences of LGBTQ+ people are rarely reported in the media, despite the fact that discrimination against sexual and gender minorities is aggravated by climate disasters (McKinnon, Gorman-Murray, and Dominey-Howes 2017). This invisibility perpetuates and compounds vulnerability: "whether or not one survived Hurricane Katrina was not simply the result of living below or above sea level; it was also about being taken into account in the city's planning for the future, being thought of as someone to consider in light of possible disasters" (Hall 2014). The vulnerability of LGBTQ+ people has been constructed through histories of political, economic, and social injustices that have rendered LGBTQ+ communities more likely to be homeless, living in poverty, chronically ill, and institutionalized in detention centers, prisons, and mental hospitals (Goldsmith, Mendez, and Raditz 2021). It is this structural oppression that creates vulnerability, which is exacerbated by invisibility.

This chapter examines the potential of collaborative documentary filmmaking to facilitate a praxis of climate justice by addressing this need for visibility and self-definition in LGBTQ+ communities. I began *Fire & Flood: Queer Resilience in the Era of Climate Change* following my own lived experience of the Tubbs fire in Santa Rosa, California, in 2017. Centering our queer commonality, I began filmed semistructured interviews with LGBTQ+ people who experienced the fires and the near-simultaneous disaster of Hurricane Maria in Puerto Rico. Two years into this project, I entered a PhD program, pushing me to examine my positionality as an insider/outsider to the primarily QTBIPOC communities I am representing. In the spring of 2020 I deepened the collaborative quality of this project through another round of interviews with the same collaborators to ask for feedback on the representations I produced in the first draft, and to seek their advice moving forward. This chapter is my reflection on this process of visibilizing queer resilience. I use reflexive autoethnography to strategically disrupt the illusion of objectivity that is key to the "powerful position of the narrator" (Adjepong 2019). I reflect on the impact of my situated knowledge as a queer climate justice activist representing my own community, as well as my positionality as an Anglophone white settler representing QTBIPOC. Throughout, I argue that collaborative documentary film offers tools for climate justice researchers inside and outside of the communities they study to move toward practices of representational justice to lift up the epistemologies of the front lines while mitigating the dangers of misrepresenting the forefronts.

Representational Justice

Scholars have long discussed the importance of recognition within environmental justice and the insistence "on being seen and heard by both a mainstream environmental movement and a government that has, for the most part, ignored them" (Schlosberg 2003). Climate justice scholarship has shown that the people who have contributed the least to climate change are experiencing the greatest impacts from its causes and consequences. No matter how many Indigenous, queer, disabled, femme, young people of color from around the world manage to attend climate conventions, their stories do not appear in the plutocracies' public records of the discourse, nor are they able to enact any power over decisions through the bureaucratic process they have been invited to witness. And so environmental justice movements have repeatedly articulated that justice must also be participatory, giving the front lines of crisis the opportunity to set laws and policies, and justice must also be representational, uplifting the cultural power of the perspectives, values, and cosmovisions of the front lines.

This need for representation persists. In *Black Faces, White Spaces: Reimagining the Relationship of African Americans to the Great Outdoors*, cultural geographer Carolyn Finney describes how in the wake of Hurricane Katrina, the two main representations of African Americans were of people helplessly stranded on their roofs and of people looting local businesses: either "victims or villains." She argues that the negative misrepresentation of African Americans, alongside the lack of representation of the diversity of African American lives outside these negative stereotypes, not only limits the way that Black people in America view themselves, but also works intertextually to reinforce and "naturalize" this representation and fuel "systemic institutional processes of racialization." Finney describes how this intertextual process creates a "regime of representation" that is "difficult to resist, dismantle, or transform." This "victims or villains" representation is compounded by the absence of representation of African Americans in nature. Finney's content analysis of forty-four issues of *Outdoor* magazine published over a ten-year period, found only 103 of 4,602 pictures containing people were of African Americans (Finney 2014, 78).

LGBTQ+ people as a community have also been invisibilized within the climate justice movement, as well as in nature itself. The examples abound, from queerness's association with urbanity, which obscures rural queer livelihoods, to the way that trans embodiments are blamed on environmental contamination. At the root of these disparate examples is the long ideological concept within Christianity that LGBTQ+ people are "unnatural" and "crimes against nature" (Mortimer-Sandilands and Erickson 2010; Gaard 1997). This ideology can be readily seen today in the declarations from evangelical leaders that blame disasters on LGBTQ+ people as God's punishment for their sin (Tierney 2019, 158). The disaster literature reveals how the history of resistance to social marginalization in the LGBTQ+ community creates the conditions for capacity and resilience during a disaster (Gorman-Murray et al. 2017). Yet queer scholars of color have also written extensively on the emergence of a "homonormative" LGBTQ+ representation, fueled predominantly by the political agendas of affluent, white, cisgender gays and lesbians who have advocated for civil rights through a platform of assimilation to the militaristic consumerism of American nationalism and collusion with settler colonialism (Puar 2006). As this homonormativity has become nearly hegemonic, it has created its own "regime of representation" that obscures the possibility that LGBTQ+ people are also poor, trans, rural, people of color, and among the front lines of vulnerability and forefronts of climate justice.

Representational justice works within the sphere of ideological struggle to combat the controlling images of hegemonic culture and replaces them with a liberatory culture based in self-representations. In a meeting in early 2019 Zephyr Elise,[3] Indigenous advisor on the film, expressed to me the need for positive representations of QTBIPOC in climate justice narratives:

> I am here in a very rural, very poor, often-forgot-about county in the western peninsula of Washington. We have one incorporated town. We have two Indigenous reservations. We have lots of Two-Spirit and queer and trans youth, including trans youth of color, and they don't often get to see themselves represented in any kind of media, let alone in stories addressing the climate crisis. . . . Our youth are hurting and struggling because they don't see themselves, they don't see that they have anything to offer. So that's why representation matters, and that's why representation for this climate crisis really matters. And not just the statistics that paint a really bleak picture of our existence, but something that shows them that they can be part of the solutions, that they have the resilience and the understandings and the wisdoms and the teachings embodied within them.

And so it is with this imperative that *Fire & Flood* centers on a narrative of resilience rather than vulnerability. While the vulnerability of LGBTQ+ people during disasters is also invisible and in need of scholarly attention, the process of visibilizing resilience disrupts "victims or villains" misrepresentations and creates a more empowering representation rooted in histories of survival and triumph over oppression.

I launched *Fire & Flood* with the desire to create a representation by and for the queer climate justice community, but I have had an uneasy relationship with my differences from my collaborators. As I sought to uplift the stories of QTBIPOC, I also removed myself from the film, as if invisibilizing my whiteness would erase the impact of my positionality. In a review of the documentary *Paris Is Burning*, bell hooks makes this explicit critique of Jennie Livingston, the white lesbian filmmaker who remains invisible: "it is easy for viewers to imagine that they are watching an ethnographic film documenting the life of Black gay 'natives' and not recognize that they are watching a work shaped and formed by a perspective and standpoint specific to Livingston. By cinematically masking this reality . . . [she] assumes an imperial overseeing position that is in no way progressive or counter-hegemonic" (hooks 2014, 151). Visual representations have the potential to naturalize difference and hide the historical, social power relations that produce difference. There is therefore both a great power and responsibility in the process of visibilization.

To interrogate my positionalities, I look to other reflexive writing in feminist methodologies, which frequently inquire into researchers' identity negotiations and performances in curating the stories of others. In one such example, Nikki Jones describes her process of navigating her racial and gender similarity to the Black women she was studying alongside the class and educational privileges that set her apart: "Who am I to tell these stories about poor, Black girls? What rights did/do I have to represent their lives? Which stories do I tell and which do I leave out?" (Jones 2009, 165). As Jones's questions evocatively highlight, the navigation of insider/outsider identities is not a harmless process; it is a matter of ethics compounded by the researcher's authority to represent and the power they build through public reception of the representations they produce. I therefore adopt Jones's questions and ask myself the same in the sections that follow:

- Who am I to tell these stories about QTBIPOC in climate disasters?
- What right do I have to represent their lives?
- Which stories do I tell and which do I leave out?

A Queer Climate Justice Activist Representing Queer Climate Justice

I still remember waking up to the smoke. In the early morning of October 7, 2017, those of us still at the Building Resilient Communities Permaculture Convergence in Northern California awoke to the smoke of thirteen fires surrounding us and the news that all major evacuation routes were blocked. As twilight approached, firefighters had finally regained enough control of Route 101 to keep it open for evacuations. For the next two weeks, all I could do was respond. I volunteered with local farmers to harvest, cook, and distribute food for evacuees. I listened to LGBTQ+ youth process the transphobia they confronted in shelters and helped them plan a welcome-home holiday party for evacuated LGBTQ+ elders (many of whom are disconnected from blood family because of homophobia). I felt the pain move through my community as a fellow organizer, a Black queer man with asthma, went into a coma from respiratory distress and now lives with irreversible brain damage. I realized that though LGBTQ+ people are hit by the same disasters, we experience them differently. Disabled, rural, low-income, homeless, incarcerated, institutionalized, QTBIPOC youth experience disaster through the kaleidoscopic intersection of all the systems of oppression that marginalize them.

The Tubbs fire in Northern California was one of several natural disasters in the fall of 2017, alongside Hurricane Maria, which ravaged Puerto Rico just two weeks prior. At the Permaculture Convergence where I had spent the days before the fires, I filmed a short fundraising video with two Boricuas who were farming in Northern California and came to the convergence to collect seeds, solar panels, and other supplies to send back to farmers on the island. Their stories of Maria shaped my lens and experience through the fires. In the following months, I interviewed people with whom I had done response work and edited them alongside the fundraising video of queer experiences in Maria. I "piloted" this short film during a workshop at the LGBTQ+ Task Force's Creating Change Conference, where attendees commented on the need for these stories to build more resilience preparedness for our communities. Over a year and a half later, I was invited by a queer farmer in Puerto Rico to make the journey to the island in person, so I bought plane tickets on a credit card and launched crowdfunding for *Fire & Flood*. I spent a week in PR, driving between a farm in the mountains, the port town of Aguadilla, and the capital, San Juan, then flew across the vast expanse of the United Settler territories, to hold another set of interviews in Northern California. Within six months, I edited all two years of interviews into a "sneak-peek" rough draft of *Fire & Flood*, which has been screened at over a dozen libraries, college campuses, and community events as part of ongoing fundraising to complete the project (see figure 8.1).

I start with this experiential origin story to uplift the "connection between experience and consciousness" addressed in Black feminist thought (Collins 2002, 24). As shown in autoethnographies of the lesbian of color canon, such as Gloria Anzaldúa's *Borderlands*, experience is an essential starting point for

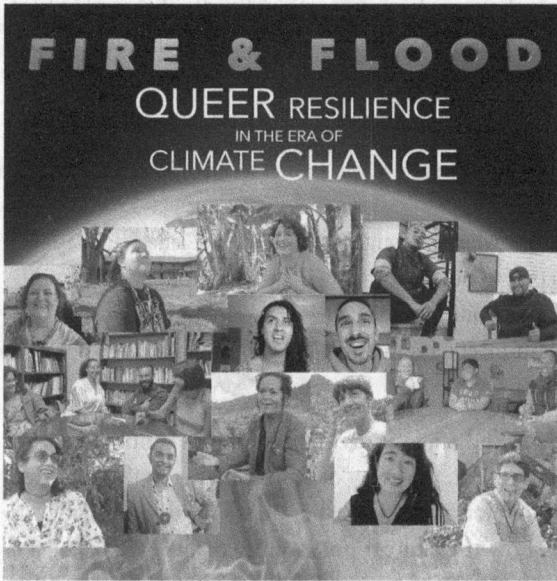

FIGURE 8.1. Cast of *Fire & Flood* collaborative documentary

grounding a project that lacks significant precedent in the literature (Anzaldúa 1987). In the midst of such extensive silence on the experiences of LGBTQ+ people through disasters, my personal experiences revealed this to me as a necessary story to tell. Beyond my queerness, it was my climate justice political positionality that had even greater impact on the film. There is a danger in epistemologies relying on "experience" alone, as "the project of making experience visible precludes analysis of the workings of this system and of its historicity" (Scott 1991, 779). Merely stating that LGBTQ+ people are more vulnerable during disasters and documenting the various forms of harassment, incarceration, and bodily harm they experience in these times does little to advance LGBTQ+ liberation struggles if the analysis is divorced from the historicity of this violence. Rather than a catalog of LGBTQ+ experiences, I wanted to use the film to visibilize a queer climate justice theoretical lens on the colonial constructions of gender and sexuality that drive the extractive economy and create social vulnerabilities. I drew on existing relationships with Two-Spirit and QTBIPOC collaborators, chose interview locations that visibilize the colonial past/present, and selected climate justice slogans as chapter titles to guide the narrative of the film. These choices were informed by my involvement in queer and climate justice activism, not from some innate essence of queer being.

This impact of my insider positionalities is particularly evident in my choices of how to represent the complex intersectionality, fluidity, and hybridity of queerness, a concept that explicitly resists singular definition. Across geographies, identities, and embodiments, the collaborators in *Fire & Flood* shared a resistance to identity labels:

c: I just like people. Gender is a construct. I don't feel like I need to identify myself.

L: Coming from the '60s when we fought so hard against labels, I find . . . [now] it's only about labels.

E: I guess I'm gender nonconforming? Labels are weird but they're fun.

T: Being queer is not just feeling something, it's about "practicing queerness."

The rejection of singular and static individual identities among so many of the collaborators requires a different way of conceptualizing the representation of queer identity, to which I turn to Jasbir Puar's writing on queer assemblage. Puar discusses how the radical potentiality of intersectionality is circumscribed by the ways in which researchers deploy it to ossify essential categories of being. She uses theory from Deleuze and Guattari to instead suggest that "bodies are unstable assemblages that cannot be seamlessly disaggregated into identity formations." How then to represent unstable bodies? This came up specifically in a follow-up interview in 2020 with a collaborator who had changed her gender and her name and underwent several gender-affirming surgeries since our first interview. We discussed how she wanted these changes represented in the film. She expressed she wanted her previous interview in the film, as well as the follow-up conversation we had about this choice. Highlighting this ephemerality destabilizes these identity categories like race, gender, and sexuality and reveals them as "events, actions, and encounters, between bodies" (Puar 2012). This idea of relationality better illuminates the self-representations that emerged in interviews.

While rejecting individual identity categories, several collaborators responded with relational and political identities. I was struck by the response from one collaborator, a disabled person of color and leader in disability justice, who answered that they identify with the queer community. This identification "with," rather than an identification "as," reflects their sense of inclusion within a movement. In an interview with graduate students at University of Puerto Rico, another collaborator describes that she chooses to identity as a lesbian not because she feels strongly about her sexual identity but because she understands lesbianism as a rejection of the heteronormative relations that cause violence against women. She identifies as a lesbian as a statement of unity with a movement imagining and building a less violent future. In a follow-up interview in 2020, another collaborator explicitly unpacks this concept. He had identified as queer in the previous interview, but from his growing connection to his Indigenous ancestry and participation in Two-Spirit community, is now critical of the concept of identity:

> I don't think Two-Spirit people before contact with settlers were like "oh we need a Two-Spirit community." You just exist in your community. Two-Spirit isn't an identity, it's how I do things. And I do it in a way that holds masculine and feminine at once in a fluid, moving relationship. So I'm only queer insofar as I am not hetero, that's as far as I can say. And it's not just about my partner, it's about all my relationships. I don't want to be in a hetero relationship with this tree. It's my ethic, it's how I show up.

To deal with the depth of complexity in these feminist and decolonial conceptualizations of queerness, I chose to dedicate close to a quarter of the screen time to introducing all of these storytellers, their differences, and their relationships. In this way, I attempt to resist a homonormative representation of "gender and sexual minorities" and instead open space for self-representations of a coalitional movement against settler-colonial, white supremacist, cisheteropatriarchy.

Through this portrayal, the film situates the vast variability of gender and sexual identity within the broader biocultural diversity of the planet, a framing that emerged primarily from my conversations with organizers with Movement Generation in California. Queer resilience is rooted in the ecological principle that diverse ecosystems are more resilient to disruption than are monocultural ones. Queer resilience valorizes difference, renaturalizes it within the vast biocultural diversity of life on the earth, and asserts that queer, trans, Two-Spirit, and other beings who question white supremacist, cisheteropatriarchal, monocultural ways of being are at the forefront of reinvigorating the creative solutions and alternative lifeways necessary in this time of disruption and collapse. This framing contributes to the work of queer ecology to highlight the vast gender and sexual diversity of nonhuman life (Mortimer-Sandilands and Erickson 2010) by adding that cultural diversity is a necessary component of biodiversity. It pushes the climate justice movement to acknowledge the role of precolonial sexual and gender diversity within traditional ecological knowledge. Just as Two-Spirit people held unique ceremonial roles in many Indigenous societies, people who embody change and transformation have a unique purpose in this time of transition.

The representations of "resilience" in *Fire & Flood* are as diverse as the collaborators themselves. The film highlights a vibrant collage of projects: trans folks created spaces of safety and joy for each other in the shelters. Fondo de Resilencia rebuilt farmers' homes in rural PR, provided reiki and acupuncture to the farmers, and fed brigade volunteers organic, locally grown meals throughout their service trips. Healing Clinic Collective and Ancestral Apothecary created free, bilingual, and accessible traditional healing clinics for the Latinx community of Sonoma County. Cuir Kitchen Brigade organized Nuyoricans to can local produce to send to community on the island. After the Kincade fire, the Disability Justice Culture Club's "Power to Live" campaign used solar-powered batteries to charge respirators, mobility aids, and refrigerated medications while protesting the privatized energy utility, PG&E. As the COVID-19 pandemic erupted, a QTBIPOC media collective in PR continued organizing mutual aid for communities displaced by the earthquakes. These wildly different examples, among many other projects, highlight the diversity of approaches to disaster recovery that build new embodied knowledges of queer ecological futures: prefigurative political acts that repair connections between land, body, mind, spirit, and community—practicing reimagined, regenerative relationships.

Since "queerness" is an unfixed, mutable, coalitional identity, my choice to name this diversity of approaches "queer" resilience is a choice of "strategic essentialism." Third-world feminist Gayatri Spivak coined this term to describe the formations of temporary essentialisms in postcolonial struggles to bring a sense of unity and visibility to a group seeking political demands (Eide 2016). Organizing this project around "queerness" offered a bridge between me, those on the front lines who share this identity, and queer audiences who may connect to climate justice through these representations. The way I chose to do this came from my standpoint—not an innate queer essence but a political positionality that I built through decades of participating in queer and climate justice political community. Yet in this "strategic essentialism" centered around queerness linger the ongoing dangers of essentialism. To tell a story of coalitional queerness requires that I equally investigate the impact of my differences from my collaborators—differences of embodiment, identity, and experience.

A White Anglophone Settler Representing People of Color

Examples abound of the way in which my Anglophone whiteness shapes the representations present in the film. In going back through my footage from Puerto Rico, I cringe at my Spanish. "Lo siento que mi español es tan feo," I would say often. I typically began in Spanish and offered collaborators the option of holding the interviews in either English or Spanish, with the disclosure that if we continued in Spanish, I would rely on the transcription later to fully understand what they shared. In the footage from one interview, S makes it very clear that she did not trust me to understand and properly translate her if we were to do the interview in Spanish. She was not the only one who chose to speak to me in English, even when Spanish was more comfortable and fluent, and rightfully so: during an interview with graduate students at the University of Puerto Rico, I asked about a topic that one of them had just addressed. I only realized my error later when reading the transcript. I tried to mitigate some of the impacts of language difference by hiring queer Puerto Ricans for Spanish transcription and translation, who were able to give specific attention to the nuances of nonbinary Spanish. All the same, the in-person language barrier deeply impacted my ability to build trust. Notably, none of the graduate students I spoke with replied to my invitation for follow-up interviews.

While the language difference is an obvious way that misrepresentation can occur, there are many other ways that my ignorance and directorial power can combine to create harmful misinterpretations. Recognizing this, I decided to adopt more of the practices of collaborative documentary film to guide me through a new phase of cocreating the next draft. Collaborative documentary film names a wide set of practices that have emerged over the past half century to facilitate a transfer of power to allow those who are typically misrepresented or unrepresented to instead create their own self-definitions that can speak back to controlling images of mainstream society (Ruby 1991, 198). This approach treats peoples as collaborators, not "interviewees." As bell hooks explains, "As objects, one's reality is defined by others. . . . As subjects, people have the right to define their own reality, establish their own identities, name their history" (hooks 1989, 42–43). In early summer 2020 I reached out to the same collaborators and invited them into another set of interviews and media sharing.

When I asked collaborators for their reflections on their representation in the first draft of the film, I specifically named my whiteness and invited reflections about the representations of race in the film. In response, Fe Román commented on how constructions of race are different in Puerto Rico: "Here in Puerto Rico we don't go around thinking 'oh I'm Latinx.' That identity, that space, comes when we have to go outside of the island. I feel like this conversation is almost imposed on us." This came up again in an exchange with Luz Cruz from Cuir Kitchen Brigade, who had helped screen the sneak peek at an event for BIPOC in food justice. Whereas white-dominated academic audiences mostly shared excitement for the diverse representation of queerness,

Luz described the BIPOC audience's disappointment "that the film was pretty white." They explained: "People on the island think about whiteness in respect to ancestral lineage. . . . Being Puerto Rican is an ethnicity, not a race, and your ancestors could've been mostly Spanish and you're still born in the island, but because of your ancestors, your race is white, and there's this dynamic that I don't think white people understand or [have] thought about fully."

These comments called my attention to not only my whiteness but also my positionality as an outsider to the island and the racial politics of Puerto Rico. The lack of representation of Black queerness in the first draft of the film came up again in two other interviews. S, who is mixed and white-passing, put it to me bluntly: "There are no visibly Black people in the film, right? I mean, it's your film, but I think it would be good to have some visibly Black people." I had been uneasy about this since my interviews in 2019, when I realized that none of the Black Boricuans whom I invited for an interview replied. Another comment from Fe highlighted how my whiteness likely contributed to distrust from potential Black queer collaborators on the island. In discussing this lack of Black representation in the film, Fe started to encourage me to reach out to a particular collective, but then gave it a second thought and said: "I know that they have this very political stand on being . . . how can I say this? The work that they do is focused on doing it 'by ourselves, for ourselves,' meaning that they might not be so interested in being interviewed." Similar to the queer strategic essentialism I employed to frame *Fire & Flood* as a representation for and by queer people in the climate movement, Fe's comment describes a desire for Black queer representations for and by other Black queer people. I strongly respect this insider media-making and defend this choice for self-representation. How then to move forward?

These conversations opened space for me to seek advice and feedback from others, and to grapple transparently with how to represent Black queerness in a way that resists tokenization. In processing this, I decided to reach out to Deseree Fontenot, a Black queer climate justice organizer whom I have known for over four years. I interviewed Desi in the fall of 2017 to give some critical framing on land justice for the Creating Change workshop, but then I did not incorporate the interview into the sneak-peek draft of *Fire & Flood* because she was not directly in the fires or response work. Now, with pandemic and additional rounds of fire broadening the scope of the film, I talked with Desi about bringing her old interview back in for the next draft. During our conversation, I directly expressed my discomfort with tokenizing her, to which she agreed that "in terms of weaving in anti-Blackness into a film about queer and trans people navigating crisis . . . its hard, because it's all connected, which we know, and also . . . if I talk about the intersection of Blackness, queerness and crisis and no one else does . . . that would be a little weird." Through discussing it more, Desi agreed to let me use her previous interview and to collaborate on the next draft. It is not yet clear how we will weave representation of Black queer climate resilience into the next draft, but the process will be through collaboration with those represented. I am grateful to these collaborators who gifted me their energy in helping me create a plan for more just representation grounded in honest and authentic relationships. While the final representation will undoubtedly be limited by my outside positionalities, I am committed to including excerpts from these conversations to bring to the surface the "difficult questions underlying what it means to be a white person in a white supremacist society creating a film" (hooks 2014, 151). I also hope it offers QTBIPOC more opportunity to launch their own projects of self-representation.

However, reflexivity alone is not enough for representational justice: I strive to materially support collaborators. Climate justice requires reparations: those who have benefited from extraction returning resources to those who have had their land and labor stolen from them through colonization, slavery, and the globalization of the extractive system driving climate change (Movement Generation 2016). While far from a structural process of reparations, fundraising is one practical way that I materially resource queer resilience. Since the day I launched my crowdfunding campaign, I have given half of what I raise directly to collaborators. These funds have helped folks through personal hardships and supported their communities through disaster response. With the other half of the funds I raise, I compensate the LGBTQ+ people advising, editing, translating, captioning, producing music, and speaking at events. In the sense that these funds predominantly come from large nonprofits, libraries, and educational institutions paying to screen sneak-peek drafts of the film, this fundraising provides a way to reallocate public resources directly into the hands of communities at the front lines of climate justice.

"The Creative Function of Difference"

Who am I to tell these stories about QTBIPOC in climate disasters? I am a queer climate activist who lived through disaster, seeking to tell a story that resists heteronormative, hegemonic ways of telling disaster narratives. I am also a white Anglophone settler fumbling through relationships across linguistic, cultural, and racial differences. The stories I am able to tell and those I cannot are shaped by my embodiment, identity, experiences, and political lens, an assemblage crafted through engaging in community organizing. What rights do

I have to represent their lives? Only the rights granted to me by my collaborators through informed consent and ongoing processes of building trusting relationships. As such, I hold a great deal of responsibility to be trustworthy in respecting those rights. Given the vast amounts of ignorance I had about PR going into this project, I am lucky that I did not cause more harm (that I know of, based on who agreed to a follow-up interview). I did not receive much critical feedback from collaborators about the way they were represented. Instead, I repeatedly heard messages such as "I trust you" and "If anything comes up, I know I can tell you." In our first follow-up call, I thanked M for continuing to collaborate over the years. M shared with me that he has appreciated "the work that the film is doing, that you're doing through the film . . . the ways you've been screening it, the places you've been screening it and just the intersections, the bridges you're trying to build." By organizing this work around the strategic essentialism of queerness, this bridge offers an opportunity for others in the majority–white-led LGBTQ+ movement to engage with a QTBIPOC-centered story that exposes them to climate justice critiques of racism, colonialism, and capitalism.

This process of collaborative documentary film has illuminated for me Audre Lorde's work on the "creative function of difference in our lives" (Lorde 2007, 111). As a descendant of colonizers and enslavers, I have much to humbly learn from the descendants of those my ancestors enslaved and colonized. Through doing this work, I also have much to teach other white queer climate justice activists, as we all attempt reconnection and repair forward and backward in time. I see this as a process of "standing with," which Kim TallBear (2014) explains "is different from speaking on behalf of. Rather, one speaks as an individual 'in concert with,' not silenced by one's inability to fully represent one's people" (4). Rather than choosing inaction in fear of imperfect representation, I seek processes that enact solidarity. There is a productive tension that comes from refusing to excise myself from this community while also committing to critically evaluating the impacts of my multiple positionalities. I am not the only one among the collaborators who had considered remaining invisible—several others had expressed concerns to me around whether they were "queer enough" or "marginalized enough" to be part of this project. My insistence on my own inclusion is an extension of my insistence that they too belong.

LGBTQ+ people are part of every community on the front lines of climate change and at the forefront of movements for collective liberation of people and planet. As Pulido and DeLara argue, scholar-activists must reclaim "radical epistemic traditions" to "craft radical imaginaries" that "move beyond the limits of the real in order to imagine the possible" (Pulido and De Lara 2018). Just as they lift up decolonial border thinking and the Black radical tradition, so too must an intersectional climate justice movement lift up the radical epistemologies of queer liberation. Yet this attention to visibilizing marginalized epistemologies needs to be done with great attention to the ethics of representational justice. Collaborative documentary film provides one approach that can help other climate researchers cocreate empowering representations of invisibilized communities, rooted in collaborators' own self-definition and direction. It is an approach that requires the humility to listen, work through shame, and change course. This chapter summary is inevitably a sanitized version of what has in practice been a messy, resource-intensive, and relationally fraught process. Yet I see it as a step toward a queer horizon. In the words of Charlene Carruthers, "we are practicing and theorizing as we go. We are seeking to eradicate oppressive systems in the world. And all our effort is worth it" (Carruthers 2018).

NOTES

1. I use the acronym QTBIPOC when referring to queer and trans people of color as a community, and I use the LGBTQ+ acronym in referring to queer and trans communities that are white-dominated, multiracial, or not racially specific. When describing individuals, I use their preferred identity terms.

2. The People's Climate March "Frontlines" banner was designed by CultureStrike, now known as the Center for Cultural Power, led by QTBIPOC artists including the executive director, Favianna Rodriguez.

3. All collaborators quoted were shown drafts of this chapter and given the choice to be included by name or by initial, or to remain anonymous.

REFERENCES

Adjepong, A. (2019). Invading ethnography: A queer of color reflexive practice. *Ethnography*, 20(1), 27–46.

Anzaldúa, G. (1987). *Borderlands: La Frontera*. Vol. 3. Aunt Lute San Francisco.

Audre Lorde Project. (2014). A Solidarity letter to the climate justice movement. *Audre Lorde Project* (blog). https://audrelordeproject.tumblr.com/post/97911207957/a-solidarity-letter-to-the-climate-justice.

Carruthers, C. (2018). *Unapologetic: A Black, Queer, and Feminist Mandate for Radical Movements*. Beacon Press.

Collins, P. H. (2002). *Black Feminist Thought: Knowledge, Consciousness, and the Politics of Empowerment*. Routledge.

Eide, E. (2016). Strategic essentialism. In N. Naples, R. C. Hoogland, M. Wickramasinghe, and W. C. A. Wong (Eds.), *The Wiley Blackwell Encyclopedia of Gender and Sexuality Studies* (pp. 2278–2280). Wiley Blackwell.

Finney, C. (2014). *Black Faces, White Spaces: Reimagining the Relationship of African Americans to the Great Outdoors*. UNC Press.

Gaard, G. (1997). Toward a queer ecofeminism. *Hypatia*, 12(1), 114–37.

Goldsmith, L., Mendez, M., and Raditz, V. (2021). Queer and present danger: Understanding the disparate impacts of disasters on LGBTQ+ communities. *Disasters*. https://doi.org/10.1111/disa.12509.

Gorman-Murray, A., Morris, S., Keppel, J., McKinnon, S., and Dominey-Howes, D. (2017). Problems and possibilities on the margins: LGBT experiences in the 2011 Queensland floods. *Gender, Place and Culture*, 24(1), 37–51.

Hall, K. Q. (2014). No failure: Climate change, radical hope, and queer crip feminist ecofuture. *Radical Philosophy Review*, 17(1), 203–225.

hooks, b. 1989. *Talking Back: Thinking Feminist, Thinking Black*. South End Press.

———. 2014. *Black Looks: Race and Representation*. Taylor & Francis.

Jones, N. (2009). *Between Good and Ghetto: African American Girls and Inner-City Violence*. Rutgers University Press.

Lorde, A. (2007). *Sister Outsider: Essays and Speeches by Audre Lorde*. 1984. Crossing.

McKinnon, S., Gorman-Murray, A., and Dominey-Howes, D. (2017). Disasters, queer narratives, and the news: How are LGBTI disaster experiences reported by the mainstream and LGBTI media? *Journal of Homosexuality*, 64(1), 122–144.

Mortimer-Sandilands, C., and Erickson, B. (2010). *Queer Ecologies: Sex, Nature, Politics, Desire*. Indiana University Press.

Movement Generation. (2016). From Banks and Tanks to Cooperation and Caring: A Strategic Framework for a Just Transition. https://movementgeneration.org/justtransition/.

Pineda, C. (2015). The dangerous erasure of queer and trans* people of color from the climate movement. *Bluestockings Magazine* (blog). May 5. http://bluestockingsmag.com/2015/05/05/the-dangerous-erasure-of-queer-and-trans-people-of-color-from-the-climate-movement/.

Puar, J. K. (2006). "Mapping U.S. homonormativities." *Gender, Place and Culture*, 13(1), 67–88.

———. 2012. "I would rather be a cyborg than a goddess": Becoming-intersectional in assemblage theory. *PhiloSOPHIA*, 2(1), 49–66.

Pulido, L., and De Lara, J. (2018). Reimagining "justice" in environmental justice: Radical ecologies, decolonial thought, and the Black radical tradition. *Environment and Planning E: Nature and Space*, 1(1–2), 76–98.

Ruby, J. (1991). Speaking for, speaking about, speaking with, or speaking alongside an anthropological and documentary dilemma. *Visual Anthropology Review*, 7(2), 50–67.

Schlosberg, D. (2003). The justice of environmental justice: Reconciling equity, recognition, and participation in a political movement. *Moral and Political Reasoning in Environmental Practice*, 77, 106.

Scott, J. W. (1991). The evidence of experience. *Critical Inquiry*, 17(4), 773–797.

TallBear, K. (2014). Standing with and speaking as faith: A feminist-indigenous approach to inquiry. *Journal of Research Practice*, 10(2).

Tierney, K. (2019). *Disasters: A Sociological Approach*. John Wiley & Sons.

PART 3

Resistance and Activism for Urban Climate Justice

CHAPTER 9

Beyond the Racial State, Racial Capitalism, and Settler Colonialism

Toward a Grassroots Climate Justice

DIEGO MARTINEZ-LUGO

Youth climate justice activists articulated ten demands, titled the "Colorado Plateau People's To-Do List," during the 2018 Uplift Climate Justice Conference that took place in the Cedro Peak Campground, outside of Albuquerque, New Mexico. These were as follows:

> Keep fossil fuels in the ground, end sacrifice zones, protect our water, stop fracking, decolonize the plateau, end systemic oppression, defend sacred sites, honor Indigenous self-determination, revive equitable economies, abolish capitalism.

Uplift, named after the geologic process that led to the rise of the red rock region, started as a small for-youth-by-youth climate justice conference in 2015 that now acts as a unifier and amplifier of the Southwest youth-led climate justice movement. Following my own scholarly activist endeavors organizing around climate justice, I was one of eleven organizers who helped organize the 2018 Uplift conference. This chapter reflects on how a particularized form of climate justice was imagined and articulated during Uplift, conceived from below at the grassroots level by a range of activists, endeavoring to disentangle climate justice from the logics of the racial state, racial capitalism, and settler colonialism. This project calls us to critically reimagine climate justice as a radical praxis and a theory of deconstruction of self-determination, anticapitalism, and decolonization.

The question guiding this chapter is: how do we enact a climate justice beyond the logics of the state and capital? In other words, when the state is the only entity that can legislate and regulate against fossil fuel emissions, and when petitioning the state via traditional democratic ideals (popular assembly, voting, and effective participation) falls gravely short of effecting the radical change needed to curb the climate crisis, what course of action should we take? In what I term grassroots climate justice, I contend that climate justice (CJ) is being renegotiated and transformed by activists, organizers, and coconspirators as a concept and a practice. Grassroots climate justice is both a lens for critiquing and unmasking CJ struggles (climate justice as deconstruction) as well as a praxis that resituates the power of change into the grass roots (climate justice in motion). When so many resources and energy go into mainstream climate policy and where significant power is concentrated, focusing on technocratic solutions and legislative avenues to effectuate change, there is a different power to be contended with; this is why I focus on grassroots climate justice. Via a critical engagement with grassroots movements and Indigenous peoples, this chapter will examine how activists construct a climate justice that rejects the racial state, take matters into their own hands, and draw the links between climate injustice and the underlying socioeconomic structures of racial capitalism and settler colonialism that prefigure and fuel climate change.

This chapter begins by exploring how climate justice has been traditionally conceptualized in geographic literature as predominantly focusing on international negotiations and Rawlsian forms of justice. I proceed by reviewing literature on the racial state, racial capitalism, and settler colonialism and advance the notion that climate justice must analyze these logics and structures as precursors to climate change. Then I continue by introducing the Uplift conference, focusing on detailing its ethos and themes, and I situate the geographic location and environmental struggles on Chaco Canyon in New Mexico. Following this introduction of climate justice, I subsequently analyze how Uplift collectively conceptualized and created a grassroots and embodied climate justice not centered on equality of harms and exposures but, alternatively, as a framework of liberation and emancipation from the systems of oppression that underpin and give rise to climate change. I investigate the panels, talks, and workshops that took place during Uplift. I conclude by introducing the concept of grassroots climate justice as a topography of addressing the root causes of climate change intended to aid climate justice activists and academics in their efforts against hegemonic climate injustices. Building on existing ideas in the academy and communities, grassroots climate justice is embedded in a longer history, as well as diverse spaces where Indigenous peoples and racial justice activists have been fundamental to laying the groundwork.

Uplift is more than a three-day conference where people convene to discuss abstract issues, deliver professional presentations, and network future working relationships. The conference itself, while an event, is more importantly a confluence of struggles joining together to amplify fights toward justice, discuss which tactics and strategies do and do not work, and assemble solidarity to connect varying fights spanning the Colorado Plateau. Uplift is the culmination of a longer process for Indigenous, People of Color, and youth activists involved in struggle and fights toward liberation. This convergence therefore offers an opportunity to garner strategic and tactical organizing skills and to cultivate relationality among and across social difference and environmental issues. I attended the conference as a participant in 2016, I led a workshop in 2017, and finally, I was one of the conference's organizers in 2018. At the heart of Uplift lies the crucial element that the struggles covered during the conference do not remain behind; the struggles continue beyond the three-day event but with more consciousness, more zeal, and newly learned strategies.

Understanding Climate Justice and the Racial State, Racial Capitalism, and Settler Colonialism

CONCEPTUALIZATIONS OF CLIMATE JUSTICE

The way in which climate justice (CJ) has been theorized and understood has traditionally been through functionalist approaches to justice, occurring in the arena of global climate change agreements and international negotiations with an emphasis on policy making and processes. Focus has revolved around normative theories of justice—such as avoiding harms and the sharing of burdens—pointing toward those most responsible for climate change as those who should bear the burden (and cost) of combating climate change (Caney 2014). In terms of the international regime, CJ has been utilized in ethical frameworks (Forsyth 2014) in climate policy articulated as procedural and compensatory of equitable burdens and distribution of costs as well as a compensatory analysis—meaning that countries least responsible and most affected by climate change, in particular underdeveloped nations and small island developing states (SIDS), get compensated for harms and damages (Klinsky and Dowlatabadi 2009).

Typologies of CJ, then, have been conceptualized around the classic Rawlsian conception of justice (Rawls 2009) as fairness. To rectify disadvantages of the social contract applied to various institutions and social issues, this liberal conception of egalitarianism is oriented around three fundamental ideas of justice: freedom, equality, and fairness. Regarding CJ, theories of justice have been organized around distributional justice (how to equally distribute the risks and burdens of climate change among nation-states) and procedural justice (ensuring fairness in the decision-making process) (Okereke 2010; Hayward 2007). Broadening notions of justice beyond fairness and equality in

CJ discourse have figured in advocating the breaking up of the north-south impasse in international climate negotiations (Parks and Roberts 2008), and scholars have attuned to political economic questions of resistance to neoliberal governance (Bond and Dorsey 2010). This has also included delineating the influence of strategies, tactics, and alliances activists foment in challenging climate injustices (Bond 2000, 2012). Lacking in these considerations, however, is a critical interrogation of climate change's underlying processes of racial capitalism and settler colonialism, as well as the preeminent role the racial state plays in allowing and incentivizing these processes.

THE RACIAL STATE

Not all of CJ has been theorized or understood as a top-down bureaucratic idea occurring in the international arena. In their research on food sovereignty and energy remunicipalization initiatives, Routledge, Cumbers, and Derickson (2018) construct a grassroots CJ activist mobilization in collaboration with state action. Arguing that another state is possible, necessary, and insufficient in the context of a changing climate, they conceive of the state as a terrain of struggle and possibility: the state for the authors is a space for contestation and engagement. Seeking alternative agendas for change, CJ activists' engagement with the state as a terrain of possibility is summarized as "with/against/beyond the state" (82). Their assertion rests on the need for the state to be reconfigured away from its neoliberal arrangements and that CJ activists must create a different set of relations resulting in a relational dynamic with the state. Even as the authors acknowledge the neoliberal state, as well as its production of difference, missing from their analysis is the conceptualization of the racial state and its interests in shaping and maintaining injustices and difference. The state (alongside capital) shapes not only climate injustices but also the racially unjust social order in the first place.

The state plays a central and changing role in creating and maintaining racial categories and difference and the inherent inequalities associated with racialization. Understandings of state formation and engagement with the state must speak to the racial dimensions of modern states in how "race is integral to the emergence, development, and transformation (conceptually, philosophically, materially) of the modern nation-state" (Goldberg 2002, 234). In regard to environmental justice (EJ) literature, Kurtz (2009), drawing from critical race theory, has shown that racism intersects with other social relations, including capitalism, to foster and respond to conditions and circumstances of racialized environmental injustice. Thus the state is a racial state, but the milieu of modern states spanning space means that "there are racial states and racist states" (Goldberg 2002, 233) and not just one unitary racial state. In the context of the United States, the state is a white supremacist one (Bell 1991).[1] Other than Baldwin's (2009) research showing that the commodification of boreal forest carbon management in Canada is tied to and relies on the racialization (and erasure) of Aboriginal peoples, there is a paucity of literature explicitly tying climate change to the racial state. However, just as critical race theorists Omi and Winant (2014) state that "the racial state does not have precise boundaries" (138), research on climate justice should be capacious to extend theories of the racial state to paradigms of climate justice.

Laura Pulido (2017) has cautioned EJ scholars to view the state as a site of opposition because it sanctions racial violence and creates the barriers of amelioration: "the state is deeply invested in not solving the environmental racism gap because it would be too costly and disruptive to industry, the larger political system, and the state itself" (529). Furthermore, Smith (2011) has shown that states are innately racially exclusionary—as well as ecologically unsustainable—by ostensibly claiming to be stewards of nature while, at the same time, controlling natural resources and permitting the expropriation and extraction of Indigenous land and natural resources. To this effect, Pellow (2017) has shown that social inequalities are reinforced by state power. Referring to political emancipation being fatally coupled to racialized state violence, Melamed (2015) articulates this incommensurability of appealing to the state as the arbiter of justice as "the state establishe[s] itself as at once the protector of freedom" in addition to the "counterviolence to the violence of race" (77). The racial state, therefore, has both a political and economic investment in maintaining climate and environmental injustices, and in turn, efforts toward climate justice defined or mediated by the state are mere concessions. If climate justice is to seek true justice (meaning a justice that does more than offer concessions and rectifies the wrongs of climate change), we must understand that we cannot abscond the racial state's entrenched investment in racial inequality and oppression. Scholars of climate justice must not only contend with the racial state but also orient themselves to processes of how racial capitalism and settler colonialism lead to climate injustices in the first place.

RACIAL CAPITALISM

The racial state, a product of society and history, exists to mediate the accumulation of capital by capitalists and that resulting class oppression that Lenin citing Marx speaks about is inherently racialized. Capitalism sanctifies the pursuit and expansion of capital at all costs—whether they are human or ecological life. "According to Marx, the state is an organ of class rule, an organ for the oppression of one class by another; it is the creation of 'order' that legalizes and perpetuates this oppression by moderating the conflict between classes" (Lenin 2015, 43). The culmination of this unfettered pursuit of growth is climate change as an externality of capitalism's extraction, accumulation, dispossession, industrial and carbon economies, and unceasing expansion. The continued plundering of Indigenous land and looting of natural resources is not color-blind. Following Melamed (2015), this is to say that "capitalism is racial capitalism" (77). Cedric Robinson in his foundational book *Black Marxism: The Making of the Black Radical Tradition* (2000) has shown that racism was a structuring logic, not a by-product, of European capitalism—which itself emerged from within feudalism, not as a negation of it. To this extent, racialism was already present in feudal relations of European peoples. As European civilization developed, the expansion of capitalist society, via the bourgeoisie and the absolutist state, pursued and relocated racial directions and ideology. Racial capitalism sought to not only hierarchize relations of Europeans to non-Europeans but also develop a capitalist society pursuing racial directions so that "as a material force, then, it could be expected that racialism would inevitably permeate the social structures emergent from capitalism" (Robinson 2000, 2). This accounts for the racial character of capitalism, one that delineates the production of social difference, the disruption of social relations, and racialized violence inherent in accumulation. Racial capitalism, then, allows for a critical engagement with the processes of invasion, colonization, settlement, imperialism, chattel slavery, and primitive accumulation (Kelley 2017)—and, I might add, the driving forces of climate injustice—that shape the world we live in today. Moreover, my assertion is that climate change and its injustices and uneven influences and impacts are a formulation of the machinations of racial capitalism over time.

Taking race, racialization, and racism seriously into environmental research is critical in understanding how the landscape of environmental inequality and degradation are formed vis-à-vis systems of racial oppression and logics of racial capitalism. In *A Billion Black Anthropocenes or None*, Kathryn Yusoff (2018) deconstructs geologic portrayals of the Anthropocene as a race-neutral epoch caused by humanity writ large. Drawing from Black feminist thought, Yusoff counters geology's knowledge-making practices and centers the Black experience under the Anthropocene to highlight how Black people have been violently subjugated and dehumanized as natural objects—read, natural resources like gold, metal, or oil in West Africa—to be objectified, categorized, and extracted. As Yusoff writes that "Black and brown death is the precondition of every Anthropocene story" (66), the zenith of her argument is that geology's logics of both the mapping and extraction of resources are founded on colonial violence. Understanding the corporeal and typological colonial violence of racialism, racism, and racialization inherent in racial capitalism reckons with the social and economic drivers of climate injustices, rather than merely diagnosing its symptoms.

Scholars have adeptly tethered the linkages of racial capitalism to climate injustice. Carmen Gonzalez (2021) has tied racial capitalism and coloniality—which Hernández (2018) understands as a matrix of power of the persistence of colonial situations and relations beyond formal colonial administrations—to climate injustice. She does this by centering a race-conscious analysis of climate change by examining the racialized nature of the extraction of fossil fuels (based on the colonization and subsequent looting of the Americas) and United Nations legal and policy responses to climate displacement and migration (to which the most vulnerable are Black and Brown people, such as those living in Small Island Developing States). In effect, Gonzalez makes climate justice more malleable by expanding it to incorporate race-conscious and decolonial narratives that examine "the cradle-to-grave impacts of carbon capitalism [which] has the potential to unite diverse and powerful social movements that reject militarism, extractivism, economic inequality and racism" (2021, 129). Adopting a four-part definition of climate injustice comprising distributive, procedural, corrective, and social injustice, Gonzalez demarcates a deficiency in this definition by exhibiting how race is inscribed in the sacrifice zones and overall history of carbon capitalism.

What racial capitalism shows us, therefore, is that everyone is vulnerable to climate change; however, vulnerability and exposure are uneven and disproportionately impact nonwhite bodies. In the United States, racism (the devaluation and objectification of human beings based on phenotype, ethnicity, indigeneity, geographic location, and/or origin) and racialization (classifying specific bodies as superior or inferior with specific traits) work to produce the logics of anti-Black and anti-Native racism drawing from what Santos (2014) has described as the "abyssal line" (70): the demarcation of those deemed fully human from those deemed less than human. The ultimate goal in analyzing the climate crisis through a lens of racial capitalism is to pave the way for the forging of alliances of diverse social groups and social movements across militarized borders and disparate landscapes.

SETTLER COLONIALISM

Just as climate justice scholars should interrogate the role of the racial state in climate injustice as state-sanctioned violence as well as racial capitalism's impetus in climate change, the *longue durée* of settler colonialism must prefigure into our analysis. Settler colonialism is a structure, not an event. As Wolfe (2006) has famously argued, it is a structural process—one that is never complete—by which an imperial power seizes and usurps Native territory to permanently eliminate and replace Indigenous peoples with a settler population and settler futurity (see also Estes 2019b). Resistance to settler colonialism, however, does not prefigure into Wolfe's widely cited framework. "Missing from the framework of the elimination of the native," says Estes, "is the emancipation of the native by the native herself" (2019a, at 23:53). Making the connection from settler colonialism to climate injustice is critical. Kyle Whyte has argued that climate injustice is not a new occurrence emerging out of a twenty-first-century understanding of emissions and tipping points, but as a cyclical history of colonialism and forced anthropogenic change (such as industrial capitalism) toward Indigenous peoples in what he terms "colonial déjà vu" (2016). Climate change, and its lexicon of injustices including pipeline development, carbon-intensive economies, and land dispossession, are not new to Indigenous peoples.

"Settler colonialism, as an ecological form of domination," writes Whyte, "is environmental violence" (2018, 137). This environmental violence has destabilized, among others, the climate system; has disrupted human relationships with the environment; and has attempted to sever Indigenous relationality. Take the Indigenous resistance against the Dakota Access Pipeline (DAPL) led by Standing Rock Sioux tribal members. Framed as issues of water quality, protecting sacred sites, or even climate justice, the #NoDAPL movement was an amalgamation of these critical factors, yet it was also a continuation of resistance against the injustice of settler colonialism and a fight for self-determination (Estes 2019b). Whyte demarcates the settler-colonial history of DAPL by outlining the perpetual violation of the 1851 Treaty of Fort Laramie, which defined territories and particular tribal group governance of 134 million acres of land, as well as the fracturing of the Great Sioux Reservation into six smaller reservations under the 1889 Sioux Bill. This history of displacement and outright subjugation of self-determination by the U.S. settler state developed the preconditions for the construction of DAPL (Whyte 2019).

Climate justice must not privilege either the racial state, racial capitalism, or settler colonialism as exclusive systems where one structure has isolated impacts or more fundamental repercussions than the other on climate injustices. Rather, CJ's conceptualization should view these three structures with distinct permutations as operating in tandem, as entangled and enmeshed, one propping up and fueling the other. This enables, first, the deconstruction of the foundation on which climate change figures to subsequently and holistically tackle the issue of climate change. Lakota historian Nick Estes (2019b) writes: "Racial capitalism was exported globally as imperialism, including to North America in the form of settler colonialism. As a result, the colonized and racialized poor are still burdened with the most harmful effects of capitalism and climate change, and this is why they are at the forefront of resistance" (28).

Settler colonialism and racial capitalism—while operating distinctively temporally and spatially, affecting racialized peoples in disparate ways whose specificity must be distinguished between—are intimately interwoven. The racial state buttresses them. Furthermore, we should be invigorated by Ranganathan and Bratman's (2019) astute claim that "climate justice is not just about climate" (18). Their analysis of an abolitionist climate justice via Washington, D.C.'s history of settler colonialism, racial slavery, and federal housing segregation depicts a genealogy that is interconnected resulting in climate injustices today—particularly in climate vulnerability and official policies of mainstream resilience thinking. This chronicling of climate change as an amalgamation of settler colonialism and racial capitalism, upheld by the racial state, illustrates climate injustices as more than emissions and rising temperatures. Doing so equips climate justice as a permutation to dissect the simultaneity of these structures and, more importantly, to envision radical action to construct alternative worlds. This dissection of the structures of the racial state, racial capitalism, and settler colonialism connects to the "possible urban futures under climate change" schematic outlined in the introduction of this book in fruitful ways. For starters, the approach elicited in this chapter directly names and explicitly confronts the social and material processes underpinning climate injustices. In terms of a secure and equitable scenario, it makes attainable alternative urban futures that promise both safety and inclusivity (see the introduction to this volume).

Uplift: Situating Climate Justice through Place

In November 2014 Uplift was founded near the Grand Canyon in northern Arizona by ten youth organizers. Their goal was to connect activists to climate justice struggles on the Colorado Plateau and to use this as a medium to spark meaningful action. The Colorado Plateau refers to the predominantly high desert region province (consisting also of deeply carved canyons, plateaus, and a scattering of conifer forests) centered on the Four Corners region of the southwestern United States, totaling an area of 240,000 square miles stretching across Arizona, Utah, Colorado, and New Mexico. This diverse geographic landscape is the ancestral and rightful home to the Diné, Hopi, Havasupai, White Mountain Apache, Kaibab Paiute, Hualapai, and Pueblo peoples. Each year Uplift connects, trains, and empowers young people (youth defined as between the ages of fifteen and twenty-five years) to act on climate justice during an immersive three-day conference where participants replace the traditional confines of ostentatious conference hotels with an explicitly place-based experience of camping out in national parks or forests. Engaging in workshops, trainings, creative spaces, and panels aimed at networking people together, Uplift amplifies the power of storytelling from frontline communities facing environmental injustices to equip young activists with the tools necessary in building meaningful relationships to support grassroots movements of resistance across the Colorado Plateau.

Uplift takes a conscientious place-based approach to its conference, rotating where the conference will be held each year, and thus space is one of the main determining factors in which themes will be covered. The same is true for which struggles, voices, and stories will be centered and amplified. The first Uplift conference took place outside Flagstaff, Arizona, in 2015, followed by Durango, Colorado, in 2016, then Moab, Utah, in 2017, and finally, for the analysis of this chapter, 2018's Uplift conference took place in northern New Mexico outside of Albuquerque. Stories of resistance to environmental injustices, racism, and extraction are amplified through a centering of place-based struggles so that no Uplift conference is the same. In 2018 Uplift concentrated on the three main themes of resistance, protecting, and healing. These guiding urgent and timely themes were placed into a spatial context with Chaco Canyon in northern New Mexico, the ceremonial center of ancestral Puebloans, which is home to sacred cultural sites and critical water resources for the Pueblo, Hopi, and Diné peoples.

Chaco Canyon lies in the arid to semiarid San Juan Basin, scattered with dry scrub piñon and juniper forests, and is an immensely vital cultural and archaeological site. Deemed a UNESCO World Heritage Site of outstanding universal value and recognized as a National Historical Cultural Park, Chaco Canyon holds an immeasurable spiritual significance for the Diné, Hopi, and Pueblo peoples. From 850 to 1250 CE, the Chacoan peoples constructed over 230 outlier settlements, considered architectural and engineering wonders, due to their massive structures and interconnected network of ancient roads (San Juan Citizens Alliance 2020). Despite the ecosystem services and sacred spiritual factors Chaco Canyon provides, over 93 percent of Chaco Canyon has been leased to oil and gas activities, primarily through hydraulic fracturing. Commonly known as fracking, this process uses horizontal drilling and the high-pressure injection of water, sand, and chemicals hazardous to human health to extract natural gas from bedrock formations.

As it stands, there are currently over thirty-seven thousand oil and gas wells with the Bureau of Land Management (BLM), relegating Chaco Canyon as a sacrifice zone with the proposal of three thousand new oil and gas wells (Frack Off Greater Chaco 2020). The BLM in 2018 alone auctioned over forty-four thousand acres of ancestral tribal lands in Chaco Canyon under an antiquated 2003 Recourse Management Plan (RMP) that does not stipulate analyses of the impacts of industrialized fracking on the environment and its inhabitants, nor does it take into account Indigenous considerations (Frack Off Greater Chaco 2020). The 1973 decision by the Nixon administration to slate the Four Corners region (which is predominantly Nation tribal land) as a national energy sacrifice zone (Horning 2017) functions as a legislative precedent permitting widespread and reckless oil, gas, coal, and uranium mining from the region. This sweeping racial capitalist expropriation of land and resources and settler-colonial disregard of tribal treaties and Indigenous self-determination is still being viscerally felt to this day.

A Grassroots Climate Justice: Movement from Below

Climate justice incorporates a diverse cluster of definitions, yet there is no universally used definition that exists; its agreed-upon definition is as elusive as the greenhouse gas emissions that fuel it. As explored earlier, most academic discourse on climate justice tends to epistemologically center and favor distributive justice (that of a "fair" distribution and outcomes of the costs and benefits of climate change) and procedural justice (group recognition and inclusion in decision-making processes), which inadvertently relegates CJ to be mediated and adjudicated by the racial state. While demands on the state for distributive and procedural justice can be important to the future of the climate justice movement—my efforts are not to completely disregard a plurality of CJ action—groups such as Uplift are challenging this state-centered struggle, precisely because the racial state is seen as complicit in the structures of settler colonialism and racial capitalist accumulation. However, as outlined earlier in the section on the racial state, avenues of radical action on the climate by appealing to the racial state are—in the eyes of the state—circumspect and apropos to the status quo. To quote Swyngedouw (2010), "we have to change radically, but within the contours of the existing state of the situation . . . so that nothing really has to change" (219).

Despite this tremendous challenge, grassroots climate justice provides a way to unsettle mainstream notions of climate justice to imagine a climate justice that is not dependent on the intransigence of the state to enact radical change. The reality is the state can and will do nothing to abolish the underlying structures of violence and exploitation that catalyze, underpin, and currently fuel the climate chaos in which we find ourselves. Patrick Bond (2012) asks CJ scholars if there are opportunities for a kind of antisystemic infused CJ movement for "movement below" against "paralysis from above" (250), and I contend that grassroots climate justice offers a strategic maneuver and course of action. Grassroots climate justice, as both a concept and a practice, is both a lens for critiquing and unmasking CJ struggles and their fountainheads (as deconstruction) and a praxis that resituates the power of change into the grass roots (as action in motion). This framework of grassroots climate justice is innately a site of resistance because it subverts status quo climate politics by confronting the climate crisis—and its roots—head-on. By rejecting piecemeal concessions, it implements its own mechanisms of subversion and imagines and constructs its own alternative and just futures. The radical activist imaginaries associated with grassroots climate justice go against the palatable avenues of climate justice, which operate within the confines of racial capitalism and the racial state's reformism, and instead construct climate justice as an embodiment, praxis, and epistemic conceptualization from below. The Uplift climate justice conference of 2018 selected the three main themes of resistance, protecting, and healing and asked the following questions:

1. How do we combat systems of oppression so that we can create a more equitable world?
2. How do we protect sacred earth and communities?
3. How do we come together to do meaningful healing work?

These guiding questions highlight that, for Uplift, climate justice is not restricted solely to the climate; rather, it embraces Kimberlé Crenshaw's (1990) notion of intersectionality, which acknowledges that inequalities and systems of oppression do not act independently of one another. Instead, they coproduce the lived conditions of the everyday. As such, Uplift envisioned climate injustice as a network of interrelated systems of oppression. One Indigenous activist stated at the conference: "we have to have meaningful conversations, we have to have hard conversations that take a lot of time. But that is part of growing a meaningful movement towards a more healthy, just, sustainable community. We need to recognize that there are points of *unity*, points of *interconnectedness* that we all share. This fight is one fight for *all people*, if we want to have a future for *all beings*" (emphasis added).

The three themes of the racial state, racial capitalism, and settler colonialism, while they play distinct and differentiated roles across space and time, are all interconnected in the interest of a grassroots climate justice. The very nature of Uplift viewed ostensibly disparate issues, such as migration and food justice, as interrelated and connected social movements to one another. In what follows, I break out salient themes from this effort into separate objects of analysis for clarity, but in practice, the issues are very much interconnected, and the boundaries separating these issues are not so neat and tidy.

RESISTING THE RACIAL STATE:
CLIMATE JUSTICE AS ACTION

Uplift organizers sought to transform climate justice from merely being a descriptive term into a practice in perpetual motion—a verb—to elicit action: "climate justice occurs when . . ." Thus, Uplift's website defines climate justice as follows:

> Climate justice is a vision of inherent respect for all life. Our interpretation of climate justice comes largely from the visions and climate justice work of frontline communities. We dream of land that is healthy, habitable for the more-than-human world, and free from colonial exploitation and ownership. Through creative grassroots organizing, we foster a livable future with clean air, clean water, renewable energy, regenerative food systems, and liberated peoples. We transcend capitalist economic systems and create new ways of being that center equitable communities, deep relationships, and a healthy earth. (Uplift Climate, 2020)

This articulation demonstrates the commitment to the agency of marginalized communities and takes aim at the exploitative institutions and structures propping up the scaffold of climate change. This conceptualization of climate justice problematizes relations to institutions, such as the state, and envisions how we can build environmentally just communities beyond the state and its apparatuses. This normative justification for climate justice is amorphous, which potentially elides "pragmatic" implementation; however, its emphasis on grassroots organizing, mutual aid, and solidarity extends the epistemological boundaries of what is possible under a radical imaginary for the future. For example, in workshops activists spoke to issues not traditionally considered climate justice, such as healing from multigenerational trauma and nurturing community accountability outside of the police state.

The Colorado Plateau People's To-Do List, the ten demands collectively agreed upon for enacting radical climate justice, was inspired by the People's

FIGURE 9.1. Colorado Plateau
People's To-Do List

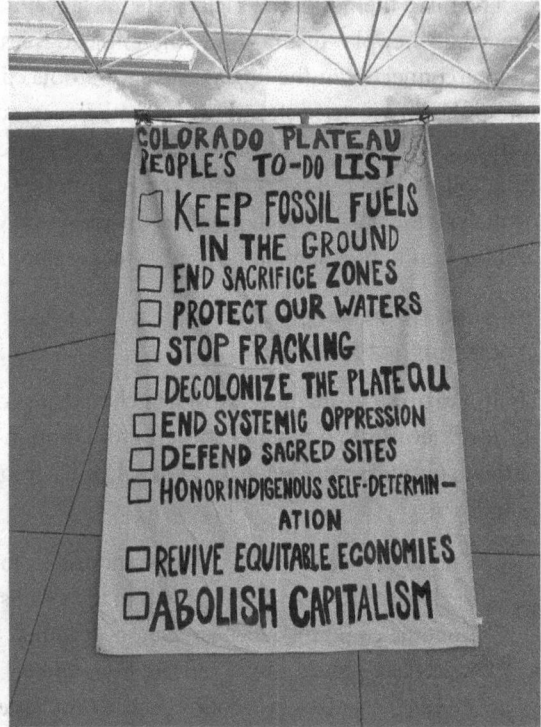

FIGURE 9.1. Colorado Plateau People's To-Do List

To-Do List, articulated by SustainUS during the twenty-second United Nations Conference of the Parties (COP) in Marrakech, Morocco, in 2016 (see figure 9.1). SustainUS is a youth-led climate justice organization that mobilizes U.S. youth to engage in domestic and international advocacy efforts, sending delegations to each COP. Originally titled the Presidential To-Do List, the list of demands started out as a protest urging COP to radically include climate justice in their agenda, but in the wake of Trump's election, "presidential" was crossed off in red and replaced with "people's." SustainUS wrote: "Our 'People's Presidential To-Do List' was never really meant for the president. It was always meant as a checklist for ourselves, a reminder of the world we, the people, will create." Inspired by SustainUS's proclamation, Uplift enacted similar thinking to go beyond the confines of the racial state to ameliorate and mitigate the root causes of climate change. While instituting some of the changes in the demands would require a long journey ahead, the impetus was to decouple CJ's locus of change from the state to the grass roots, by the people and for the people. In other words, grassroots climate justice functions as a promulgation to radically imagine other avenues of instituting fundamental climate action. This is exemplified, for example, by the second demand in bold black letters to "end sacrifice zones," which are construed and maintained by the racial state.

EXTRACTION AND RACIAL CAPITALISM:
CLIMATE JUSTICE AND DECONSTRUCTION

Uplift also recognized the racial character of capitalism during the panel "Beyond Extractive Economies," whereby panel activists spoke to energy and resource extraction on Indigenous Nations, as well as strategies and visions to move beyond extractive economies. An Indigenous activist with Tó Nizhóní Ání (Sacred Water Speaks), a community organization focused on the protection of water sources on Black Mesa in Navajo Country, spoke to mining extraction of coal, oil, gas, and uranium on Black Mesa. While the Peabody Energy–owned Kayenta Mine that powered the Navajo Generating Station powered Phoenix,[2] many Diné and Hopi people who live closest to the mine do not have access to that energy, nor do they have running water (Larsen 2018). To make matters worse, at the time of 2018 Uplift's conference, 99 percent of the workforce at the Kayenta mine are Indigenous (Johnson 2018). The prevailing discourse at Uplift coupled the threats of climate change to long histories of systemic oppression and past injustices: as the Uplift coordinator discussed, "when we see [I]ndigenous peoples, people of color, poor people dying or displaced from climate chaos, we have to understand that the disproportionate impacts of climate change that oppressed peoples face today aren't accidental, but rather a result of societal inequity" (Larsen 2018).

Just transitions have been lauded as one tactic to end reliance on fossil fuels; however, a racial capitalist perspective deconstructs the power relations of who owns and produces the energy, and whose land it is on needs to be taken into consideration. During the same panel, when the topic of a just transition from an extractive economy to a sustainable economy came up, an Indigenous anarchist interrogated whether Indigenous communities even benefit from green measures. In cautioning against green capitalist measures and their inherent power relations, he questioned, "Does that change the power relationships of exploitation, of the land and the people? I think we need to think beyond that." Viewing racial capitalism as a focal point views the means of production (in this case energy production) in tandem with who is doing the laboring, which means that these power relations must be transformed—not just the source of energy production. "I'm an abolitionist," stated the Indigenous anarchist, "looking historically at slavery. I wouldn't want a more just transition from bad slavery to a lighter, better, reformed version of it. I want a clean break, because we shouldn't have slavery to begin with." Will there continue to be dispossession, the expropriation of land, accumulation, severe inequalities, and racism across a just transition? Uplift radically rethought what this all meant under racial capitalism.

SETTLER COLONIALISM'S PERPETUATION UNDER CLIMATE CHANGE

During an educational workshop on fossil fuel extraction on Navajo land, a young Indigenous leader with Tó Nizhóní Ání spoke to the relationship between energy extraction and forced relocation, displacement, and the impounding of sheep for coal mining. The young Indigenous activist poignantly evoked the history of the Long Walk of the Navajo, the forced exiling and ethnic cleansing of more than ten thousand Navajo and Apache peoples who were forcibly relocated from Arizona to New Mexico in the 1860s (Denetdale 2009). Due to increasing energy needs of the United States in the 1960s and the discovery of coal and uranium on the Navajo Nation, the U.S. government imposed both a building moratorium and livestock reduction for Navajos living on Hopi Partition Land. This action resulted in the 1974 Navajo-Hopi Settlement Act, which created an artificial boundary between the two tribes. Approximately ten to fifteen thousand Navajos who found themselves on Hopi land were once again forcibly relocated (Lacerenza 1988). During the panel, the young Indigenous activist drew juxtapositions of forced relocation under settler colonialism of the past to that of the present: "the fourth relocation will be from climate change."

As the three-day conference progressed, nestled in a piñon and juniper forest, close to a hundred participants had to walk to workshops, panels, or keynotes by traversing the campground. This created a sense of community, not only socially but also ecologically, where participants inhabited the environment and Indigenous land they were professing to protect and defend. Workshops ranged from Indigenous-rooted direct-action training, to anarchist community organizing for disaster preparedness, to the discussion of tools, knowledge, and strategies to dismantle systems of oppression for Uplifters. Panels covered various Indigenous groups resisting the proliferation of fracking in the Greater Chaco Region, migrant justice amid climate crisis, food justice in the Southwest, and moving beyond extractive economies. Time and again, Uplifters articulated settler colonialism as a structure that needed to be resisted in addition to organizing against climate change.

On the concluding day of the conference, the Uplifters left the New Mexico National Park and carried out a carefully planned demonstration in the middle of downtown Albuquerque, in support of the Frack Off Chaco movement. Youth activists, some attending their first ever demonstration, carried colorful protest banners and chanted in unison, and they gave speeches on diverse issues. An activist concluded her speech with a quote from Aboriginal elder Lilla Watson: "If you have come to help me, you are wasting your time. If you have come because your liberation is bound up with mine, then let us work together." The multiplicity of hopes and demands for a grassroots climate justice from below expanded climate justice to mean interracial and transnational solidarity by way of prioritizing Indigenous self-determination and decolonization.

By Way of Conclusion

After three uplifting days of Indigenous-led storytelling and workshops, I saw that the concept of climate justice is currently being transformed by activists into both a lens for seeing and critiquing struggles and a praxis that resituates the power of change into the grass roots, into what I have termed grassroots climate justice. This sees grassroots climate justice as a plurality, where many articulations of justice—that of the radical, decolonial, and anticapitalist—can be found. Viewing the state as an adversary, one of the central machinations of racial capitalism, Uplift grounded climate justice in community organizing and forged a critical engagement with Indigenous struggles and social movements to subvert the systems of power that credit hegemonic avenues for climate action. In turn, this resituated the power of change in the hands of Indigenous peoples, young people, and People of Color to fight for climate justice in and from the grass roots. This emphasizes the vital role of agency of marginalized populations of not simply being targets of domination but agents of change who resist their subordination and imagine and create more just worlds. Overall, the framework of grassroots climate justice can be expanded to other climate justice struggles in various geographies and locales. Its tenets can equip activists and academics alike with a critical lexicon of deconstruction and empowering movement from below, by the grassroots. Grassroots climate justice can be connected both conceptually (as theorizing) and materially (as praxis) to actuate a climate justice that fits the needs of the most vulnerable. Uplift's invoking the processes of the racial state, racial capitalism, and settler colonialism serves as analytics for climate justice to dissect, deconstruct, and unmask the systemic culprits of climate change.

Uplift has always been embedded in a broader set of actions, institutions, and relationships, and as of this writing, Uplift continues. Uplift broke off from their nonprofit funders and became fully grassroots in 2019. By no longer being beholden to what is deemed appropriate or acceptable by funders, Uplift has become more unwavering in their principles and political beliefs. The Uplift conference carried forward in 2019 in Gallup, New Mexico, with the theme of reclaiming narratives through community, justice, and action. Due to the COVID-19 pandemic, the 2020 conference moved to a digital format, was renamed Uplift Digital Convergence, and shifted its resources to support COVID-related mutual aid networks in the Colorado Plateau. Inspired by the 2020 uprisings for racial justice, Uplift began a webinar series on abolition connecting climate justice to police, prison, and immigration detention abolition. Uplift moved away from the place-based nature of hosting the conference in a community of struggle, but their adaptation to the pandemic was out of their control. Finally, the 2021 Uplift Digital Convergence was free and virtual, titled "Rooted Communities, Stronger Together" to incorporate workshops on mutual aid, abolition, and environmental justice. Uplift has and continues to sustain grassroots climate justice along and beyond its purview as a three-day conference, extending to the struggles activists face when they return home.

NOTES

1. See also Coates (2014): "Black nationalists have always perceived something unmentionable about America that integrationists dare not acknowledge—that white supremacy is not merely the work of hotheaded demagogues, or a matter of false consciousness, but a force so fundamental to America that it is difficult to imagine the country without it."

2. After Uplift 2018 took place, citing the falling costs of natural gas and renewable energy, the Navajo Generating Station closed on November 18, 2019. The closure's impact on jobs and economic security disproportionately impacted Navajo Nation and Hopi Tribe peoples, who were once the vast majority of the workforce. For more information see https://www.greentechmedia.com/articles/read/navajo-generating-station-coal-plant -closes-renewables.

REFERENCES

Baldwin, A. (2009). Carbon nullius and racial rule: Race, nature, and the cultural politics of forest carbon in Canada. *Antipode*, 41(2), 231–255.

Bell, D. (1991). Racial realism. *Connecticut Law Review*, 24, 363.

Bond, P. (2010). Climate justice politics across space and scale. *Human Geography*, 3(2), 49–62.

———. (2012). *Politics of Climate Justice: Paralysis Above, Movement Below*. University of Kwa Zulu Natal Press.

———. (2015). Can climate activists' "movement below" transcend negotiators "paralysis above"?. *Journal of World-Systems Research*, 21(2), 250–269.

Bond, P., and Dorsey, M. K. (2010). Anatomies of environmental knowledge and resistance: Diverse climate justice movements and waning eco-neoliberalism. *Journal of Australian Political Economy*, 66, 286.

Caney, S. (2014). Two kinds of climate justice: Avoiding harm and sharing burdens. *Journal of Political Philosophy*, 22(2), 125–149.

Coates, T. (2014). "The Case for Reparations." *Atlantic*, June. https://www.theatlantic.com/magazine/archive/2014/06/the-case-for-reparations/361631/.

Crenshaw, K. (1990). Mapping the margins: Intersectionality, identity politics, and violence against women of color. *Stanford Law Review*, 43, 1241.

Denetdale, J. (2009). *The Long Walk: The Forced Navajo Exile*. Infobase.

Estes, N. (2019a). Abolishing Columbus and Indigenous resistance. *Red Nation Podcast*, October 23. https://directory.libsyn.com/episode/index/show/therednation/id/11710970.

———. (2019b). *Our History Is the Future: Standing Rock versus the Dakota Access Pipeline, and the Long Tradition of Indigenous Resistance*. Verso.

Fisher, S. (2015). The emerging geographies of climate justice. *Geographical Journal*, 181(1), 73–82.

Forsyth, T. (2014). Climate justice is not just ice. *Geoforum*, 54, 230–232.

Frack Off Greater Chaco. (2020). *Greater Chaco Coalition Primer*. https://www.frackoffchaco.org/the-issues.

Goldberg, D. T. (2002). Racial states. In D. T. Goldberg and J. Solomos (Eds.), *A Companion to Racial and Ethnic Studies* (pp. 233–258). Blackwell.

Gonzalez, C. G. (2021). Racial capitalism, climate justice, and climate displacement. *Oñati Socio-Legal Series*, 11(1), 108–147.

Hayward, Tim. (2007). Human rights versus emissions rights: Climate justice and the equitable distribution of ecological space. *Ethics and International Affairs*, 21(4), 431–450.

Hernández, R. D. (2018). *Coloniality of the U.S./Mexico Border: Power, Violence, and the Decolonial Imperative*. University of Arizona Press.

Horning, J. (2017). Reader view: Crisis is opportunity for Four Corners region. *Santa Fe New Mexican*, May 6. www.santafenewmexican.com/opinion/my_view/reader-view-crisis-is-opportunity-for-four-corners-region/article_4565692d-38ab-52e3-a6a6-dbab360a22be.html.

Johnson, G. (2018). Uplifting communities on the Colorado Plateau in the fight for climate justice. *Medium*, October 28. medium.com/uplift-climate/uplifting-communities-on-the-colorado-plateau-in-the-fight-for-climate-justice-6677c5203578.

Kelley, R. D. G. (2017). What did Cedric Robinson mean by racial capitalism?. *Boston Review*, January 12. https://bostonreview.net/articles/robin-d-g-kelley-introduction-race-capitalism-justice/.

Klinsky, S., and Dowlatabadi, H. (2009). Conceptualizations of justice in climate policy. *Climate Policy*, 9(1), 88–108.

Kurtz, H. E. (2009). Acknowledging the racial state: An agenda for environmental justice research. *Antipode*, 41(4), 684–704.

Lacerenza, Deborah. (1988). An historical overview of the Navajo relocation. *Cultural Survival*, September. www.culturalsurvival.org/publications/cultural-survival-quarterly/historical-overview-navajo-relocation.

Larsen, B. (2018). In the desert Southwest, young leaders are reimagining what a climate conference can be. *Pacific Standard*, September 25. psmag.com/environment/in -the-desert-southwest-young-leaders-are-reimagining-climate-conferences.

Lenin, V. I.(2015). *State and Revolution*. With a new introduction by T. Chretien. Haymarket Books.

Melamed, J. (2015). Racial capitalism. *Critical Ethnic Studies*, 1(1), 76–85.

Okereke, C. (2010). Climate justice and the international regime. *Wiley Interdisciplinary Reviews: Climate Change*, 1(3), 462–474.

Omi, M., and Winant, H. (2014). *Racial Formation in the United States*. Routledge.

Parks, B. C., and Roberts, J. T. (2008). Inequality and the global climate regime: Breaking the north-south impasse. *Cambridge Review of International Affairs*, 21(4), 621–648.

Pellow, D. N. (2017). *What Is Critical Environmental Justice?*. John Wiley & Sons.

Pulido, L. (2017). Geographies of race and ethnicity II: Environmental racism, racial capitalism, and state-sanctioned violence. *Progress in Human Geography*, 41(4), 524–533.

Ranganathan, M., and Bratman, E. (2019). From urban resilience to abolitionist climate justice in Washington, D.C. *Antipode*, 53(1), 115–137.

Rawls, J. (2009). *A Theory of Justice*. Harvard University Press.

Robinson, C. J. (2000). *Black Marxism: The Making of the Black Radical Tradition*. University of North Carolina Press.

Routledge, P., Cumbers, A., and Derickson, K. D. (2018). States of just transition: Realising climate justice through and against the state. *Geoforum*, 88, 78–86.

San Juan Citizens Alliance. (2020). SJCA Guide to Greater Chaco. http://www.sanjuan citizens.org/chaco.

Santos, B. S. (2014). *Epistemologies of the South: Justice against Epistemicide*. Paradigm.

Smith, M. (2011). *Against Ecological Sovereignty: Ethics, Biopolitics, and Saving the Natural World*. University of Minnesota Press.

SustainUS. (2016). Post election within COP22. *Medium*, November 9. medium.com /sustainus/post-election-within-cop22-d28b7465bc5a.

Swyngedouw, Erik. (2010). Apocalypse forever?. *Theory, Culture and Society*, 27(2–3), 213–232.

Uplift Climate. (2020). Our Story. https://upliftclimate.org/our-story.

Whyte, K. P. (2016). Is it colonial déjà vu? Indigenous peoples and climate injustice. In J. Adamson and M. Davis (Eds.), *Humanities for the Environment* (pp. 102–119). Routledge.

———. (2018). Settler colonialism, ecology, and environmental injustice. *Environment and Society*, 9(1), 125–144.

(2019). The Dakota Access Pipeline, environmental injustice, and U.S. settler colonialism. In C. Miller and J. Crane (Eds.), *The Nature of Hope: Grassroots Organizing, Environmental Justice, and Political Change* (pp. 320–337). University of Colorado Press.

Wolfe, P. (2006). Settler Colonialism and the Elimination of the Native. *Journal of Genocide Research*, 8(4), 387–409.

Yusoff, K. (2018). *A Billion Black Anthropocenes or None*. University of Minnesota Press.

CHAPTER 10

"Accounting" for Climate Justice

Fiscal Fights over Climate-Changed Urban Futures

SAVANNAH COX

On December 11, 2018, members of the Miami Climate Alliance (MCA), a consortium of grassroots climate and social justice organizations across Miami-Dade County, held an emergency conference call. The call was prompted by a just-published *Miami Herald* article on how the city of Miami planned to spend the first tranche of the Miami Forever Bond (MFB), a first-of-its-kind $400 million general obligation bond meant to finance climate-resilient infrastructure in the city. To the consortium's dismay, the initial round of projects had little to do with what the MCA had advocated for. For those on the call, the *Herald* piece signaled the possible erasure of hard-fought political victories, such as $100 million allocated for affordable housing and meaningful participation in bond project selection and allocation processes among historically marginalized communities.

During the call, MCA members discussed possible means of intervention. Members quickly turned to the issue of metrics—specifically the forms of expertise and evaluative techniques that the city used to prioritize bond projects. "Was this based on physical vulnerability or financial impact?" one member asked. "And who's producing these assessments?" Another member "[worried] that [the city is] going to make decisions based on millage [rates] . . . in lieu of [participatory] process."[1] How, those on the call wondered, could they publicly contest these techniques and forms of assessment—and, as a consequence, push the city to advance the more progressive political agenda of the MCA?

While these questions continue to preoccupy Miami Climate Alliance members, they also speak to the initially puzzling political significance that these activists have given to urban resilience finance and market devices such as municipal bonds. Why is this puzzling? For one, these activists are doing what many critical scholars of urban climate governance say cannot be done. They are treating technocratic accounting methods and market-oriented governance tools—so characteristic of Swyngedouw's (2009) "post-political city"—as potent political instruments in their efforts to challenge and resist the reproduction of Miami's racialized political-economic status quo as climates change. This leaves us with at least two pressing questions if we are to learn from activist movements typically seen as "on the margins" (Rice, Levenda, and Long, this volume). First, why and how is it that MCA members have seized on municipal debt as a key site of political claims-making? Second, how might the case of the Miami Forever Bond affirm and call into question the "critical conventional wisdom" on the role of markets and market devices, such as municipal bonds, in the pursuit (or negation) of the climate-just city (Collier 2011, 9)?

In this chapter, I advance two related arguments. First, the Miami Forever Bond, by its very design, convenes competing imaginaries of "resilient" Miami. A dominant imaginary, held by bond and city officials, ties the bond and its related projects to imagined futures of economic prosperity as climates change. An alternative imaginary, held by the MCA, links the bond and associated projects to a resilient future wherein long-standing, highly racialized inequalities in the city are substantively addressed (Connolly 2014; Grove, Cox, and Barnett 2020). As I show, MCA members have used the bond to pursue what Nancy Fraser (2005) calls the "how" of justice: problematizing the rules that determine who counts (and does not count) as part of a given political community. The MCA does so through two techniques: reconfiguring expertise and resilience. Second, and following Knuth's suggestion (in this volume) that public finance is too often treated as a post-political area of governance, I argue that devices like the Miami Forever Bond should be analyzed as dynamic political sites on which the (un)even distributions of climate change risks will increasingly be (re)imagined, operationalized, and resisted (Cox 2022). It is important to note at the outset that neither argument marks an attempt to "redeem" market-oriented climate governance. Rather, and following the lead of climate activist thought detailed here, the goal of this chapter is to better understand how market devices can be (re)made to resist unjust, climate-changed futures—as well as their limits in advancing climate-just urban futures.

Markets and Climate (In)justice in the City

Critical urban scholars have long been concerned with the role of markets and market logics in shaping urban climate governance. Building on Harvey's (1989) theories of urban entrepreneurialism, geographers have suggested that contemporary urban climate governance efforts are "framed" by neoliberal practices, such as market-oriented interventions (Whitehead 2013, 1348). In a foundational article on the subject, Hodson and Marvin (2009, 200) argue that cities around the world are increasingly working with international firms to position themselves as "the context for action" on climate change. Their reasoning is straightforward. Where the onset of climate change can render cities economically vulnerable to disaster, it can also form the basis of comparative urban economic advantage. Through concerted efforts and strategic partnerships to demonstrate their "resilience" to climate change, city governments can make themselves more attractive to private investment while also helping spur the construction of new global markets and modes of capital accumulation. For Fainstein (2015), city governments are thus likely to invest in resilience projects that secure the urban economy and its supporting infrastructures, and abstain from investing in projects that primarily benefit the poor. As city governments continue to work with risk management experts in drafting their resilience plans, urban climate governance may be "rendered technical," which hollows out the possibility of political debate and weakens government commitments to social justice and equity (Leitner et al. 2018, 1282).

Urban resilience finance, such as insurance and bond instruments that municipalities use to pay for adaptive infrastructure like raised roads, has recently emerged as a chief object of urban scholarly inquiry. This interest comes in part because urban resilience finance poses a significant practical problem to city governments themselves. Indeed, the massive projected costs of climate resilience measures—to the tune of hundreds of billions per year globally—have encouraged cash-strapped city governments to finance climate projects through bonds (Puig et al. 2016). In a nutshell, U.S. municipalities issue general obligation or revenue bonds to finance basic goods and services, such as public hospitals. Bond investors make money on the interest rate of the bonds, which are backed by an attached revenue stream or the issuer's taxing authority (Cestau et al. 2019).

Beyond their appearance as a seeming "solution" to the problem of financing resilience, these instruments are of interest because they purportedly reproduce and exacerbate inequality. Suggesting that environmental risk management in New York City and Cape Town, South Africa, is increasingly mediated by financial logics and concerns, Bigger and Millington (2020, 1, 6) argue that the use of municipal debt for adaptation and resilience both reinscribes existing urban inequalities and displaces further financial and environmental risks onto populations least able to bear them, and that "no amount of earmarked debt for adaptation will fundamentally reshape broader urban

riskscapes in egalitarian ways." Such views are consistent with common critical geographical perspectives on financialization, which suggest that "investments are diverted to those that deliver financial results rather than those that
benefit local communities" (Aalbers 2019, 10).

However, the treatment of urban resilience finance as a monolithic object
that always acts in a particular way leaves us with few analytical tools and little
conceptual guidance to make sense of the political significance that the Miami
Climate Alliance has attached to the Miami Forever Bond, as well as the significant political victories that the MCA has garnered for low-income communities of color in Miami through the bond. These accounts do not help explain
how a financial instrument such as a municipal bond can both (re)produce inequality in cities and be (re)made to resist the reproduction of such inequalities. If we are to learn from activists "at the margins," one of the goals of this
edited volume, we must give both perspectives equal consideration (Rice, Levenda, and Long, this volume).

Actor-network theory–inspired (ANT) work on market formation, or what
Çalışkan and Callon (2010) call marketization, helps us grasp the nitty-gritty
"how" that we are tasked with considering: that is, how markets are pieced together and thus how markets, their formation, and their effects can be resisted.
For those taking a marketization approach, markets and their effects should
not be taken for granted. Instead, markets and constitutive market devices—
defined as anything that intervenes in the construction or operations of a market, such as a bond in a bond market—are best seen as practical, mutable, and
temporary achievements (Callon 2007). This is because the construction of
these devices requires a significant amount of material, political, and technical
investments, none of whose efficacy is guaranteed. As Fields (2018) reminds
us in her account of the construction of the post–financial crisis single-family
rental asset class, the provisional stabilization of these investments is precisely
what "[allows] financializing projects to come to fruition" (2). Their disarray
or destabilization is what "causes [financializing projects] to fail." Conceptually, then, a market takes the form of a sociotechnical *agencement*: a combination of statements and material, technical devices that can act in certain ways
pending how they are put together (Deleuze and Guattari 1988).

What does it mean to think of markets this way? For one, it reminds us that
to study actually existing markets and market devices is to study politics, as
conflicts, disputes, and relationships of domination are always at work within
them (Knuth, this volume; Li 2007). When used as an analytic method, which
I do here, treating market devices as assemblages also means that we must
attend to their messy, contested construction, always keeping our eyes open

to the spaces and processes in which "markets, and market rule, can be disrupted" and reconfigured (Fields 2018, 5). In attending to the construction of the Miami Forever Bond, I also pay attention to the shared visions of desirable "resilient" futures underlying MCA and Miami government thought and action on the kind of work the bond should do, and how they attempt to realize these visions in practice. These shared visions—which help shape and are shaped by a given technology, such as a municipal bond—are what Jasanoff (2015a, 2015b) calls sociotechnical imaginaries. Multiple imaginaries can exist within a given entity. Through practices of embedding—transmitting ideas about ideal futures into durable institutions, practices, and materials—imaginaries work to "condition and constrain" normative visions of (un)just futures (Jasanoff 2015a, 14). Imaginaries are often competing: powerful institutions such as legislatures may work to "[elevate] some imagined futures above others," just as alternative futures can be advocated for and embedded "from below the seats of power" (Jasanoff 2015a, 4; 2015b, 323). Tracing these imaginaries and their conflictive interactions can help account for the power imbalances and the unique, value-laden roles of human imagination that ANT-inspired approaches to the study of contemporary problems sometimes miss.

In what follows, I examine the contested construction of the Miami Forever Bond and the competing imaginaries of resilience that MCA officials attempt to embed in bond practices. This analysis is based on eighteen interviews with MCA members, financial experts, and city officials who have worked on the bond; six weeks of fieldwork consisting of attendance at MCA and partner organization meetings; and content analysis of relevant documents (e.g., bond documents, publicly available correspondence between the MCA and the City of Miami government, and MCA bond outreach and campaign strategies, which MCA members shared with me on request).

"We're Showing the World That We're Doing Something!"

For city officials involved in the construction of the Miami Forever Bond, the resilient future that the bond will help realize is one of sustained investment and growing tax bases as sea levels rise. As one resilience official told me, while he and his colleagues "live in the little world of Miami," the "financial world is changing [its practices] on climate risk." For this expert, the bond "[shows] the [financial] world that we're doing something," which can "change how we are evaluated by investors . . . and make us the investee of choice" as climates change.[2] Importantly, the "little world" of Miami—that is, the everyday experiences and concerns of residents—barely appears in expert imaginaries. Describing his work as a form of storytelling, a senior bond manager told me that the bond is about "our image . . . first and foremost on a global level [to] the people in insurance, finance, investment. Then national, where we've gotten a lot of bad press. And then in the state legislature." After a pregnant pause, the manager finally added, "and I guess, you know, local" (personal communication, June 14, 2019).

But the "financial world" does more than constitute the ideal bond audience for city officials. The supposed concerns of this world also help frame and legitimate subsequent bond-financed resilience interventions. Indeed, following their work with financial risk management actors to determine the costs and benefits of various interventions, Miami city officials charged with drafting and conceptualizing the bond prioritized securing vital infrastructures from future physical vulnerabilities over "secondary" concerns such as existing social vulnerabilities within the city. While stating that issues of gentrification and affordable housing are as important as the economic risks of climate change, one senior city official told me that "if we don't deal with these [economic] risks, we'll never get to these other issues [like gentrification]" (personal communication, July 19, 2018).

In many ways, the events and concerns that led to the initial construction of the Miami Forever Bond correspond well with the claims of critical urban scholars detailed in the prior section. City officials imagine that the bond will allow them to demonstrate their resilience to a "global" market audience, which they believe can form the basis of comparative economic advantage as climates change. Financial logics, whose authority climate change only seems to elevate, have inched ever further into municipal governmental thought on and imaginations of the "resilient" public good: the reproduction of the political economic status quo—but with stormwater pumps (Wakefield 2019). How is it, then, that MCA members succeeded in making parts of the Miami Forever Bond address the "little world" of Miami and its racialized political economy?

Resisting the "Financial World"

The apparent "economic necessity" of the bond failed to push the bond or the dominant imaginations of "resilient" Miami undergirding it beyond debate. Instead, the flashy, $400 million market device quickly became a contested object of local politics. The meaning of "resilient" Miami, and quite literally how it ought to be built, was still up for grabs. But first, a brief note on the mechanics of general obligation bonds in Miami. There, before a bond can appear for a general vote, a majority of the five city commissioners must approve it. For better or worse, this means that commissioners, each of whom represents different geographic districts of Miami, can push for bond funds and projects that are attached to needs and concerns in their own districts. But, as one MCA member who spearheaded the bond outreach told me, this also presents an opportunity: these commissioners are easily swayed by public opinion.[3]

In commissioner meetings about the Miami Forever Bond, which took place in the summer months of 2017, the financial instrument served as a site to address two ongoing problems and grievances among local Miamians, which have ostensibly little to do with climate change: excess taxation and affordable housing. When the commission vote took place mid-July, one commissioner dubbed the bond "Taxing Miami Forever" and promised to campaign against the bond if it made it to a general vote. This perspective received the support of area labor union leaders, who called debt issuance for sea level rise "hysteria" and suggested the city instead devote bond funds to union back payments and benefits after the city's recession-era austerity measures were overturned that year (Smiley 2017). Commissioner Keon Hardemon, who represents District Five—home to some of the poorest and majority-Black neighborhoods in Miami, such as Liberty City and Overtown—suggested that the biggest existential threat the city faces is not climate change but the affordable housing crisis. Indeed, Miami ranks seventh in the world in terms of unaffordable housing—above London and New York—which experts tie to high housing costs and dramatic, racialized income inequality (Cox and Pavletich 2019). "Every community is not going to be receptive to sea level rise and flood prevention," Hardemon said. "That is not their issue. They'll look at you and say, 'What are you talking about?'" In exchange for a yes vote, Hardemon successfully demanded that the bond devote $100 million to affordable housing, because he was "tired of hearing that poor people will no longer be able to live here [in Miami]." A final commissioner cast the deciding vote in favor of the bond, even though he said "voters . . . are not going to understand it" (Smiley 2017).

Once the bond made it out of initial legislative uncertainty, it entered the realm of general politics, thereby generating and responding to a new set of debates and actors that could (re)shape the bond's fate. The Miami Climate Alliance was one of the most significant actors involved: as the chief resilience officer of Miami told me, the bond would not have passed if the MCA had not supported it (personal communication, June 14, 2019). The organization's interest in the bond is in many ways unsurprising, given its origins in local budgets. Per one member's account, while organizing the Miami People's Climate

March in 2015, participants "discovered $0 allocated toward climate action (or mention of climate change) in Miami-Dade County's budget" (Adefris 2018). Alarmed at this absence, the MCA mobilized hundreds of residents to attend budget hearings and demand that Miami-Dade allocate more taxpayer dollars toward addressing climate change. Two important results of their efforts were the creation of the county's Office of Resilience and the first-ever chief resilience officer.

The MCA's prior experience with Miami government budgets also helps explain their initial skepticism of the bond—precisely because they believed that ongoing problems in the "little world" of Miami, such as persistent, racialized inequality and affordable housing crises, didn't factor into bond officials' imaginations of resilient Miami. For the MCA members I interviewed, the bond as initially presented to voters was about shoring up bond ratings and insurance premiums, and speaking to "people who don't actually live here [in Miami]" (personal communication, October 11, 2018, June 26, 2019). Nevertheless, they suggested that the funding for resilience—or at least for their visions of resilient Miami—was "necessary" and that the bond "was a good opportunity to change bureaucratic systems that didn't really work for people." For MCA members, then, the bond could only become "both farsighted and just" if they succeeded at embedding their own resilience imaginaries into it (Adefris 2018). These imaginaries chiefly consist of the substantive addressing of highly racialized inequalities through the systematic (re)allocation of resources to historically marginalized communities of color and (re)valuation of the voices of these communities such that they are equal to those of their wealthy, white, and propertied counterparts (see figure 10.1). For the MCA, the material changes involved in this resilience imaginary are of urgent importance: because long-standing practices of Jim Crow segregation have confined poor people of color to renting in highly elevated parts of the city, they are particularly susceptible to displacement as the city increasingly attaches higher-elevation luxury development to its resilience efforts (Ariza 2020). Two key embedding techniques—reconfiguring expertise and resilience—were important in the MCA's work.

One crucial way in which MCA members worked to embed their own resilience imaginaries into the bond was by challenging the city's understanding of who and what counts as an "expert." As Grove, Cox, and Barnett (2020) detail, the valorization of technocratic expertise has long served to exclude minority—and particularly Black—participation in substantive political decision-making processes in the Greater Miami region. Over time, this has had the effect of unevenly distributing the costs and benefits of the region's

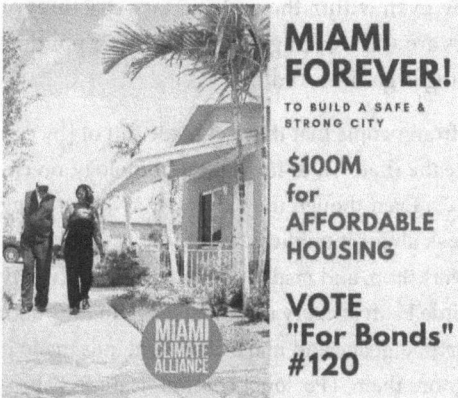

MIAMI
FOREVER!

TO BUILD A SAFE &
STRONG CITY

$100M
for
AFFORDABLE
HOUSING

VOTE
"For Bonds"
#120

MIAMI
CLIMATE
ALLIANCE

FIGURE 10.1. Miami Climate Alliance advertisement supporting Miami Forever!

economic development, typically across racial lines (Stepick et al. 2003). Many MCA members said that the bond would likely yield similar results if they did not intervene. Thus, they worked to reconfigure expertise on two fronts.

First, they advocated for the creation of a citizens' oversight board, which would oversee and provide guidance on bond project selection and allocation processes. For MCA members, a racially and socioeconomically diverse board would help "elevate voices of vulnerable populations" and ensure that their concerns are taken into account (communication between Miami Climate Alliance and city officials, 2018, in author's possession). Following weeks of protracted politicking with the city government, the MCA succeeded in altering board qualifications to incorporate community leadership as a form of expertise and to require that board members reside in the city of Miami and did not stand to financially gain from bond projects. They also "stacked the deck" and appointed three of their own members on the seven-member board (personal communication, July 17, 2018). As a result, this MCA member said, "we got a citizens' oversight board that's not just made up of white males with PhDs. This way, we can do what we can to make sure that these funds will be used in a way that benefits community members, not just downtown, not just people who invest in Miami" (personal communication, February 7, 2019).

Second, the MCA ensured that members who had participated in MCA affiliate-held climate trainings—multiweek workshops meant to make low-resource Miami locals "experts" in climate science basics—attended, and spoke at, public meetings related to the bond. This second tactic effectively concedes that local knowledge and community leadership are, in spite of their addition as board member qualifications, not enough on their own to make authoritative political claims about what resilience efforts the bond should prioritize. But it also underscores how, even within the exclusionary strictures of technical expertise, MCA members are still working to create sites for political debate. As one MCA climate training organizer told me:

> We hear over and over again from people that they get made fun of [by city government] if they don't have the rhetoric, or the right terminology, on climate change and adaptation. . . . Even though they know what's happening on their terms, they won't speak about it publicly out of fear of embarrassment or humiliation. So, we workshop, and made sure we got at least twenty people from Overtown and Little Haiti at every meeting related to the bond. Having them speak the [expert] language and make clear demands made it impossible for [the city] to ignore them. (Personal communication, February 7, 2019)

But this embedding technique also, and purposefully, takes on a public performative dimension. As another MCA member summarized, "we throw [the city] off and startle them by showing up" (personal communication, July 17, 2018). Indeed, beyond invocation of expert language, the physical presence of activist-experts at bond and budget hearings makes them and their imaginations of resilience impossible for the city to ignore.

In addition to mobilizing expert publics at bond meetings, MCA organizers reconfigured the meaning of "resilience" so that it would resonate with residents long excluded from political decision-making in the "little world" of Miami. Describing their digital communication strategy with low-income racial minorities in Overtown and Little Haiti, one MCA member said that they "framed the bond and resilience in terms of things people care about—we basically said with the bond, we'd get housing, equity, and real representation in government" (personal communication, July 17, 2018). Activists indicated that this tactic was essential to their campaign's success. "Once we conveyed to our grassroots people that we were getting pieces of equity in the bond," an MCA member told me, "there was a big rallying cry behind it" (personal communication, June 28, 2018).

In November 2017 the Miami Forever Bond passed with a 55 percent majority (Isbell 2017). Since then, and despite the routine struggles that the Miami Climate Alliance faces with the city of Miami (some of which I detailed in the opening vignette), the MCA has seen further political victories. Its members and allies in historically excluded neighborhoods have gained seats within the citizens' oversight board, which those involved have suggested is politically meaningful. Since its passage, the bond has helped fund affordable housing construction and public transportation expansion in the low-income, racial minority neighborhoods of Liberty City, Little Havana, and Allapattah (Kallergis 2019).

Conclusion

It is easy to dismiss these successes as piecemeal. In fact, many MCA members would tend to agree, having called these successes only the beginning of a long "war" with the city (personal communication, July 17, 2018). As one MCA member who helped spearhead the bond grassroots outreach strategy told me, "There's a real power struggle around whether money is just going to shoring up infrastructure and leaving our very large, vulnerable working-class population without the assistance they need to be prepared for storms, extreme heat, food security, or climate gentrification." But, and importantly for the argument of this chapter, he added that his organization believes that "the Miami Forever Bond is something that can address these things." The bond, and others like it, will remain significant sites in MCA efforts to resist the reproduction of an unequal political economic status quo as climates change while further embedding their own resilience imaginaries into the political economic machinery of the city.

But, as I have also attempted to demonstrate, it is not just any local vision of a resilient future that the MCA has sought to elevate in its ongoing political struggles with the city. Instead, it is a resilient future where low-income racial minorities who have historically been denied meaningful voice in political decision-making processes can author(ize) the forms of resilience that the city takes, and thereby advance a more just climate urbanism from below (Grove, Barnett, and Cox 2020; Silver, this volume). Thus, inasmuch as the MCA fight is about municipal finance, it is also about what Nancy Fraser (2005) calls misrepresentation, an injustice that occurs when political boundaries are wrongly framed to exclude specific individuals and communities from authoring justice claims. For Fraser, the issue of the frame—that is, the decision rules and processes that determine who or what "counts" (or does not count) as a member of a given political community—is the "central question of justice" (88) in a world whose challenges are increasingly global in scope. This is because the frame structures who is (and is not) recognized as a subject of justice, and to whom resources are (and are not) justly distributed. That the MCA is problematizing the frame through a municipal bond—supposedly impossible in a highly financialized, "post-political" city such as Miami—makes their work even more important for scholars, activists, and practitioners seeking to confront the climate crisis in just, equitable, and creative ways.

But even if Miami resilience finance remains a battle of competing imaginaries and is not yet a post-political fait accompli, it is important to note that the players and the imaginaries they seek to advance through the bond are not equal. As I detailed in the opening vignette, activists are attempting to advance alternative futures through the language and procedures of municipal finance—and against a deeply rooted sociotechnical imaginary that has historically tied visions of the "good life" in Miami with rapid, racialized economic growth (Connolly 2014). This imaginary, I have tried to show, is still at work in city official thought on the Miami Forever Bond, and by accepting the terms of municipal finance, MCA activists are in some ways dependent on them. Over time, the sustained usage of financial expert knowledge among MCA members may have the effect of reinforcing expert authority and technocracy, thereby having the somewhat paradoxical effect of destabilizing the very alternative imaginaries that MCA members are still working to stabilize vis-à-vis the Miami Forever Bond (Kim 2015). Thus, while financial instruments such as municipal bonds will likely be increasingly prevalent in urban resilience efforts, they should still be treated as one site among many in which activists should engage city planners and policy makers in their fights for urban climate justice.

I have also attempted to show that assuming that urban resilience finance only follows one path risks passing over its status as a site of intense political contestation over the meaning of resilience and the concrete form(s) that it takes. In my case, urban resilience finance serves as such a site through the very technicalities of the Miami Forever Bond process: convening the practices, concerns, and imaginations of (extra-)local actors who design, evaluate, and purchase pieces of the bond with those of residents who may (or may not) vote to pass the bond in the first place, and who may have very different understandings of the "good" resilient life and how it is best achieved.

This argument is not of purely academic importance. If it is the case that, at least in the United States, urban resilience continues to be financed in part through debt issuance—and to date, there is no indication otherwise—it is of utmost importance to continue to treat debt, and circulations of finance more broadly, as intensely political fields of struggle in collective fights for a more just climate urbanism (Knuth, this volume; Silver, this volume). Likewise, scholars might avoid the urge to read the tools and techniques of market-oriented urban governance as devoid of politics and the possibility of political space. This too is an imaginary of sorts, and one whose analytical dominance among researchers seems only to grow as climates change. It is worth taking seriously. But if we are to follow the lead of activists described here, we should still try to resist it.

NOTES

1. Millage rates represent the amount per every $1,000 of a property's assessed value, and are used to calculate local property taxes. This MCA member's emphasis on millage rates speaks to a broader MCA concern that Miami government will prioritize exchange value (e.g., property values) over collective deliberation about other non-exchange forms of value (such as culture or community, per the MCA member's statement) in bond allocations.

2. Before bond-financed projects can begin, the bond must be rated, underwritten, and invested in. These processes necessarily bring together many actors from the "financial world" (e.g., credit rating agencies, [re]insurance companies, and investment firms) whose changing practices Miami officials described here are concerned about. With respect to promotion, in efforts to attract further "global" attention to the bond, the mayor of Miami penned a *New York Times* op-ed with former U.N. secretary-general Ban-Ki Moon about the bond and the important "resilience" work it would do. In separate interviews with city officials, it was confirmed that the op-ed was meant to draw global attention to the fact that Miami is a "resilience" leader.

3. In describing his bond advocacy with one commissioner, this MCA members said, "We [the MCA] worked hard to get him there [his elected commission seat] and he knows how he got there, and so he's very much beholden to our platforms and issues . . . so we put a lot of pressure on him to get behind the bond, and it worked" (personal communication, July 17, 2018).

REFERENCES

Aalbers, M. B. (2019). Financial geographies of real estate and the city: A literature review. *Financial Geography Working Paper*, 21, 1–46.

Adefris, Z. (2018). Making the "Miami Forever Bond" a Model for Equitable Climate Adaptation. Urban Resilience Project. https://medium.com/@UrbanResilience/making-the-miami-forever-bond-a-model-for-equitable-climate-adaptation-c4826b4b209c.

Ariza, M. (2020). *Disposable City: Miami's Future on the Shores of Climate Catastrophe.* Bold Type Books.

Bigger, P., and Millington, N. (2020). Getting soaked? Climate crisis, adaptation finance, and racialized austerity. *Environment and Planning E: Nature and Space,* 3(3), 601–623.

Çalışkan, K., and Callon, M. (2010). Economization, part 2: A research programme for the study of markets. *Economy and Society,* 39(1), 1–32.

Callon, M. (2007). What does it mean to say that economics is performative?. In D. MacKenzie, F. Muniesa, and L. Siu (Eds.), *Do Economists Make Markets? On the Performativity of Economics* (pp. 311–357). Princeton University Press.

Cestau, D., Hollifield, B., Li, D., and Schürhoff, N. (2019). Municipal bond markets. *Annual Review of Financial Economics,* 11, 65–84.

Collier, S. J. (2011). *Post-Soviet Social: Neoliberalism, Social Modernity, Biopolitics.* Princeton University Press.

Connolly, N. D. B. (2014). *A World More Concrete: Real Estate and the Remaking of Jim Crow South Florida.* University of Chicago Press.

Cox, S. (2022). Inscriptions of resilience: Credit ratings and the government of climate risk in Greater Miami, Florida. *Environment and Planning A: Economy and Space,* 54(2), 295–310.

Cox, W., and Pavletich, H. (2019). 15th Annual Demographia International Housing Affordability Survey. Demographia. http://www.demographia.com/dhi.pdf.

Deleuze, G., and Guattari, F. (1988). *A Thousand Plateaus: Capitalism and Schizophrenia.* Bloomsbury.

Fainstein, S. (2015). Resilience and justice. *International Journal of Urban and Regional Research,* 39(1), 157–167.

Fields, D. (2018). Constructing a new asset class: Property-led financial accumulation after the crisis. *Economic Geography,* 94(2), 118–140.

Fraser, N. (2005). Reframing global justice. *New Left Review,* 36, 69.

Grove, K., Barnett, A., and Cox, S. (2020). Designing justice? Race and the limits of recognition in greater Miami resilience planning. *Geoforum,* 117, 134–143.

Grove, K., Cox, S., and Barnett, A. (2020). Racializing resilience: Assemblage, critique, and contested futures in Greater Miami resilience planning. *Annals of the American Association of Geographers,* 110(5), 1613–1630.

Harvey, D. (1989). From managerialism to entrepreneurialism: The transformation in urban governance in late capitalism. *Geografiska Annaler: Series B, Human Geography,* 71(1), 3–17.

Hodson, M., and Marvin, S. (2009). "Urban ecological security": A new urban paradigm?. *International Journal of Urban and Regional Research,* 33(1), 193–215.

Isbell, M. (2017). Miami Forever—The Politics behind a City's Preparation for Climate Change. MCI Maps. http://mcimaps.com/miami-forever-the-politics-behind-a-citys-preparation-for-climate-change/.

Jasanoff, S. (2015a). Future imperfect: Science, technology, and the imaginations of modernity. In S. Jasanoff and S.-H. Kim (Eds.), *Dreamscapes of Modernity: Sociotechnical Imaginaries and the Fabrication of Power* (pp. 1–33). University of Chicago Press.

———. (2015b). Imagined and invented worlds. In S. Jasanoff and S.-H. Kim (Eds.), *Dreamscapes of Modernity: Sociotechnical Imaginaries and the Fabrication of Power* (pp. 321–342). University of Chicago Press.

Kallergis, K. (2019). Miami approves $7M in bond money for affordable housing. *The Real Deal.* https://therealdeal.com/miami/2019/06/28/miami-approves-7m-in-bond-money-for-affordable-housing/.

Kim, S.-H. (2015). Social movements and contested sociotechnical imaginaries in South Korea. In S. Jasanoff and S.-H. Kim (Eds.), *Dreamscapes of Modernity: Sociotechnical Imaginaries and the Fabrication of Power* (pp. 152–173). University of Chicago Press.

Leitner, H., Sheppard, E., Webber, S., and Colven, E. (2018). Globalizing urban resilience. *Urban Geography,* 39(8), 1276–1284.

Li, T. M. (2007). Practices of assemblage and community forest management. *Economy and Society,* 36(2), 263–293.

Puig, D., Olhoff, A., Bee, S., Dickson, B., and Alverson, K. (2016). The Adaptation Finance Gap Report. United Nations Environment Programme, Nairobi. https://unepdtu.org/wp-content/uploads/2019/04/adaptation-finance-gap-report-2016.pdf.

Sassen, S., and Portes, A. (1993). Miami: A new global city?. *Contemporary Sociology*, 22(4), 471–477.

Smiley, D. (2017). Miami Commission sends $400 million general obligation bond to the voters. *Miami Herald*, July 27.

Stepick, A., Grenier, G., Castro, M., and Dunn, M. (2003). *This Land Is Our Land: Immigrants and Power in Miami*. University of California Press.

Swyngedouw, E. (2009). The antinomies of the postpolitical city: In search of a democratic politics of environmental production. *International Journal of Urban and Regional Research*, 33(3), 601–620.

Wakefield, S. (2019). Miami Beach forever? Urbanism in the back loop. *Geoforum*, 107, 34–44.

Whitehead, M. (2013). Neoliberal urban environmentalism and the adaptive city: Towards a critical urban theory and climate change. *Urban Studies*, 50(7), 1348–1367.

CHAPTER 11

Love in the Time of Climate Crisis

Climate Justice through a
Universalism of the Oppressed

ANKIT KUMAR

Dipesh Chakrabarty (2012a) suggests that in the context of the Anthropocene one needs to see human agency at two scales: as a political force and as a geological force. As a political force, humans seek justice as rights-bearing citizens, while knowing that full justice cannot be had, and therefore, humans stay engaged in survival politics. As a geological force, humans are a collective author of actions. These human images create a dilemma because climate change demands thinking human agency simultaneously at two, seemingly incompatible, scales (Chakrabarty 2012a).

In this baffling question of scale, our understanding of climate change is based on a paleo scale that traces planetary history, whereas our understanding of climate action is based on human history, its calculations and risks (Chakrabarty 2014, 9). How do we think through these two scales simultaneously to respond to the climate crisis? According to Chakrabarty (11), it is important to realize that the often misplaced history of climate change is not exclusively a history of humans (i.e., of capitalism—although capitalism, racialized and related processes of colonialism are deeply implicated). Rather, it is an amalgam of short-term human history and longer-term planetary history. The literature on the science of climate change, by presenting "new questions of scale" (astronomical, geological, evolutionary), adds a critical corollary to our "completely homocentric narrative" (Chakrabarty 2015, 154).

Following this question of scale, Chakrabarty (2015), makes a distinction between "homocentric and zoecentric views of the world" (142). This distinction is critical. By forgetting that the planet has a longer history than humans, we consistently equate humanity and planetary (as in planet earth), that is, "a zoecentric view is passed over in favour of a homocentric one" (165). The homo is the "figure of one-but-divided humanity" (159). Thinking about the planet—zoecentric—rather than just humanity—homocentric—is what

Chakrabarty calls an "epochal consciousness" (183). Epochal consciousness opens the path to think around human politics while keeping the space of politics open. This might open ways to "resist climate injustice and create alternative futures" (Rice, Levenda, and Long, this volume). How do we achieve this? How do we think zoecentrically and take responsibility as one humanity while also keeping the space of homocentric politics open?

This speculative chapter looks toward postcolonial, decolonial, and anticolonial literatures in an attempt to take a first step toward thinking zoecentrically. As in the case of climate change mitigation, adaptation to the climate crisis depends on shedding old affiliations and opening ourselves to unexpected friendship and hospitality toward others. If mitigation is understood as a "common cause," then climate adaptation also needs to be taken as a "common cause." This requires thinking through radical utopian politics. Radical utopian politics opens the space for new ideas and voices in a climate-changed world, voices and ideas that a regressive, deadlocked politics might judge as immature. Regressive politics also seeks to focus on "others" who seek hospitality in Europe and ignores acknowledging the hospitality, love, and friendship offered by these "others" now and at various critical junctures in the history. Regressive politics seeks to be othering and dominating, rather than inclusive and collaborative.

I argue that to develop a zoecentric politics, that is, a life-centric politics, the first step is equity for those who have been degraded and discriminated against as lesser human life forms—lives that colonial and contemporary evidence tells us have mattered less, and groups whose behavior and characteristics are racialized and ridiculed as being closer to nature, closer to undisciplined life. This is a reminder that Black lives do matter, and that Black lives must matter in any thinking around the climate crisis. In the Indian context, this also translates into Dalit and Adivasi Lives Matter. The first step to get to a politics where all life on earth is a politics of Black life on earth. I propose that to initiate such a politics we need to look toward these groups and excavate a universalism of the oppressed.

The rest of the chapter progresses as follows. First, I explain why this chapter looks at the colonized as interlocutors in this debate. Second, using Leela Gandhi's ideas on radical (utopian) socialism, it presents a conceptual background for what a political community premised on friendship, hospitality, and love might look like and why this is important. From this, I argue for an ethical commitment toward love for strange guests and friends. In the realm of (urban) climate justice, I put forward some examples of guests and hosts to illustrate these ideas and think through what a universalism of the oppressed could be. These examples revolve around the journey a Dutch man undertook in an electric car from the Netherlands to Australia to demonstrate the possibility of a low-carbon world.

Colonized as Interlocutors

Mediated by postcolonial, anticolonial, and decolonial thought, I look toward the colonized as interlocutors. There are two reasons for this. First, thinking zoecentrically is thinking life-centrally, that is, progressing a politics that thinks of all life on earth. I propose that a first step toward thinking of all life is centering human lives that have historically been considered as "less than human." Groups that have historically been variously essentialized as wild, disorganized, dangerous, and unpredictable, on the one hand, and innocent, needing care and guidance, on the other. Those who needed to be disciplined and regulated. In other words, groups whose position was thought to be closer to nature. In an attempt to center these lives, the chapter excavates the position of one such group: the colonized. Ashis Nandy (2009) explains how the colonial gaze came to frame the colonized as less than human. Nandy discusses the centrality of ideals of masculinity in legitimizing colonization though the doctrine of progress. Related to this was the idea of "the child" who was an "inferior version of maturity, less productive and ethical, and badly contaminated by the playful, irresponsible and spontaneous aspects of human nature" (Nandy 2009, 15). Colonialism picked this idea of "the child" from Western Europe and planted it in the colonies: "what was childlikeness of the child and childishness of immature adults now also became the lovable and unlovable savagery" of the colonized societies (16). The colonized then emerge as childlike and the less-than-humans that European humanism and universalism "worked/works to uplift."

Second, Chakrabarty (2018) outlines that it is "urgent that we devote our collective attention to questions of humanism and to the problem of how to think about and relate to . . . cultural and historical differences," so we can find togetherness without our "differences being made into excuses for domination" (145–146). This is a departure from the European universalism thought that ended/ends up legitimizing domination and colonization. I look toward the colonized for such humanism because, to rephrase Nandy (2009), between the dominator and the dominated (or elite and the subaltern), one must look to the dominated as they embody "a highest-order cognition" that inevitably embraces the dominator as a human while the dominator's reasoning excludes the dominated, "except as a 'thing'" (xv–xvi). This is an explicit politi-

cal stance that says if indeed we need to unite as one humanity and develop a
universal politics of climate change, such politics can only emerge from "the
universalism of the oppressed," "a universalism that will never be able to fold
completely into itself... any particular that enriches it" (Chakrabarty 2018,
157–158).

A Community of Epochal Consciousness:
Toward a Politics of Hospitality, Empathy, and Love

When faced with a climate crisis that will have uneven impacts, is a universal
position that guides "competitive and conflicted actions by humans" fathom-
able (Chakrabarty 2015, 142)? Leela Gandhi (2005) provides an answer. While
Chakrabarty's epochal consciousness puts the ethical responsibility on humans as a whole, Gandhi's noncommunitarian communitarianism, based on
singularity rather than individuality, can channel this responsibility. Singularity is a condition that brings together those with "irreducible difference" (26).
Such community, which rejects closure and affirms opening, is always in becoming. Risk posed by the climate crisis often carries "a profound affirmation
of relationality and collectivity" and opens possibilities for such a community
(32). Gandhi suggests that radical utopian socialism can open the path to such
collectivity. Being barred from "the drama of mainstream politics" and the inability to chime with the "established political vocabulary" is what gives utopian socialism its revolutionary spirit, its potential for alternatives (178). I propose that this pathway for coming together with irreducible difference opens
us up to thinking zoecentrically.

Within the realm of utopian socialism, premised on friendship, "the political" is a space of community that is open and hospitable—a space that offers
a reversal against social exclusions (Gandhi 2005, 19). While the state cannot
imagine or tolerate a "community without affirming an identity," friendship
resists such exclusions and creates a co-belonging that transcends identities
(26). Taking a further step, Gandhi makes a case for "philoxenia, or a love for
guests, strangers, and foreigners": an ethical commitment to "strange friends"
that establishes itself on an opposition to the state's radical exclusivity. Such
philoxenia has the potential to unfold "alternative climate futures" that transcend violence, injustice, inequalities, and climate apartheid (Rice, Levenda,
and Long, this volume). Where might we find such philoxenia, and how might
we develop it?

Gandhi (2005) proposes that such politics of friendship can develop
through a departure from the existing and a "political opening to otherness"

(183). The "other" here is a relationality that renders "politics into a performance of strange alliance, unlikely kinships, and impossible identification": a politics of philoxenia. In doing so, it reveals, and at the same time refuses, "the crisis of nonrelation" (184). What we need, then, is to imagine with "what is fragmentary and episodic" because the fragmentary does not imagine being part of the whole called the state and can therefore express ideas and knowledges that are detached from "the will that produces the state" (Chakrabarty 2012b, 274).

Mediated by these ideas of friendship, hospitality, and love for others, I provide some historical, literary, and contemporary examples to explore a case of love for all life on earth and love for strange guests. In these cases, I stand in the corner of the colonized, the subaltern actors who have historically been seen as "less than human" and "closer to nature." Thinking with Nandy (2009) and flipping the colonial politics, could one argue that taking the vantage points of the "lovable and unlovable savagery" (15) of the colonized, are our first steps toward thinking zoecentricity? Could those Others whose behavior was/is thought to be closer to the nature/undisciplined life, through their utopian thinking, be the bearers of a universalism of the oppressed, one that unites but, unlike the enlightenment universalism, does not oppress?

Love for Guest: Radical Friendship and Hospitality

Let us turn to some examples of a politics of philoxenia, friendship, and hospitality. One summer morning in 2019, I was listening to the BBC World News podcast. This particular episode ended with the story of a Dutch man, Wiebe Wakker, who had just finished traveling on land from Europe to Australia in an electric car. On his website he explains his journey as a "purpose-driven adventure with the aim to inspire, educate and accelerate the transition to a zero carbon future." He traveled with little money, relying on the hospitality and kindness of people he encountered on the way. He frames his idea, called "Plug Me In," as a "collaboration between people." The BBC anchor asked Mr. Wakker if, in this long journey, he had to sleep outside on many occasions. His reply, that in Europe there were a few occasions when he had to sleep outside or in his car, but in Asia he always found a place to stay, piqued my interest in the politics of hospitality and friendship that enabled his low-carbon journey. He explained the hospitality received in Asia as curiosity of people: "wherever the car stopped, people gathered to look, and someone offered help" (paraphrased from memory). However, I wondered if there was more to this.

His journey needed three main things: electricity to charge the car, food, and shelter. Through his website and social media, he looked for people who could provide these. On the same website he provided updates on his journey in the form of written blogs and short videos. Through these I tracked his journey across India. In total, he stayed at twenty-four places in India, mostly urban areas of various sizes and a few villages. Some of these stays were preplanned, but many were arranged on the go (often one host calling someone else to arrange the next stay) or serendipitously after reaching these places. There are three things to note. First, many people opened their homes and hearths to a stranger with whom they had limited, if any, shared identity—someone who might qualify as a "strange guest." Second, often he needed to stay in one place for a longer period than initially planned, which the hosts invariably accommodated. Third, most people did not have specific provisions—household infrastructure or high-power electricity connections—to handle heavy loads in their households. They helped charge the EV while it sometimes jeopardized their own electric infrastructure.

History is replete with such stories of travel and hospitality. One related story of travel in the other direction is that of the Indian nationalist, philosopher, and Nobel laurate Rabindranath Tagore's visit to the United States, recounted and analyzed by Dipesh Chakrabarty (2018). During his visit to Chicago in the year 1913, Tagore found accommodation with one Mrs. Harriet Moody, a wealthy Chicago businesswoman and socialite.[1] Mrs. Moody, who had a superior sociocultural position, extended hospitality and love to her guest, Tagore. This was premised on their historical and social differences. However, as they spent time in each other's company, a longer-term relationship grew. Chakrabarty sees this relationship as friendship, but "not a friendship between two people who were equal or similar in any obvious way" (96). In addition, Tagore was famously overreliant on people around him for his daily activities, which often made being around him difficult. So how did this friendship work? Chakrabarty, working with Derrida, explains that friendship involves a judgment to love: "it is more worthwhile to love than to be loved." "To love," then, signifies a higher position, one that does not seek utility and is devoid of ego (101). When there is no expectation of reciprocation, love can be inclusive of "nonhumans" too.[2]

Tagore, Chakrabarty explains, saw Moody as a mother, one who loves unconditionally—one who carried out her hospitality, friendship, and love, despite his difficult persona. While Chakrabarty focuses on the civilizational exchange—a thirst to learn from the other—as a key reason for Moody's unconditional hospitality,[3] one cannot but ask if this unconditional love and generosity also emerged from the host's historical, social, and economic position as superior to the guest. Moody had the resources at hand to provide her guest what he needed while not being "put out" herself. Contrast this with the story of our Dutch traveler. The historical and social differences flip. The hosts are often in an inferior historical, social, and economic position. What extends the hospitality then?

Gandhi (2005) posits that "the ethical agency of the host-friend relies precisely on her capacity to leave herself open . . . to the risk of radical insufficiency" (31). Is this what happens as Mr. Wakker travels through India? Many working-class people accommodate him, sharing their limited resources of space, food, and electricity. They stay open to the possibilities of his longer and unexpected stay. They risk connecting heavier loads to their electricity connection, which means that in the short term their electricity connection could break down, and in the longer term they might receive a bigger electricity bill. An incident from May 10, 2017, is telling. Mr. Wakker reached a host but the electricity current in the household was unstable. They attempted to charge at 15A (ampere), but the electricity supply was inadequate. They went down to 13A, but it did not work. They tried 10A with no success. Finally, they found that 8A was the maximum allowed voltage through this connection. They left the car connected through the night but found out in the morning that due to frequent power cuts, only 21 percent of the battery had charged. Mr. Wakker then looked for other places to charge but was largely met with disappointment, although not a lack of desire to help: "Everybody was very willing to cooperate but nowhere there was stable current." Even with inadequate provisions, many tried to help, risking their own survival. Finally, they found a small garage that had adequate current, and they were able to charge the car.

On other occasions, his car broke down more than once, and many people helped, either transporting it or him from one place to another. In a video from June 12, 2017, when his car suddenly broke down on the road, he remarked: "In India, you never have to worry that you face problems alone. Even in smallest villages people pop out and help you with your problem."

Although such help and hospitality are in no way a uniquely Indian characteristic, it is useful to reflect on it through the Indian context.[4] In Sanskrit the word for guest, *atithi*, if broken apart (*a-tithi*) means "no date." It represents someone who has no fixed arrival or departure date. The host must remain open to sudden arrivals and the possibilities of longer stays. Sawhney (2018) explains that *atithi* is a "strange and untimely figure" who commands respect by claiming a home and resources that are not hers because it is a reminder of our own "infinitely more profound un-belonging" (4). More deeply, it is a re-

minder of that ultimate uninvited, but inevitable, and often untimely guest—
death—that reminds us of our own morality. Hindi poet Jai Shankar Prasad's
poem "atithi" gives an idea of the opening that a guest demands.

> The cavern of the heart was empty,
> the house vacant.
> Let me move in quickly,
> the spirit grown to double.
> One guest came—
> I didn't recognize him.
> His feet made no sound—
> I was not aware.
> Then what happiness—
> someone in my house.
> My heart rejoiced
> that it was occupied.
> They call him "love"—
> oh yes, I knew him now.
> When his claws left their rough marks,
> then I recognized him.
> But the guest stayed on,
> didn't leave the house;
> began to play his game—
> oh! it was a lion. (Rubin and Prasad 1979)

This conflation of love and guest in Prasad's poetry is useful in excavating
the position of the guest. Prasad's lover-guest comes unannounced, yet he re-
joices in their presence. This lover-guest stays, and even though it pains the
host while inhabiting his "inside," the host accepts this as he knows that "love
is capable of wounding the heart" (Pachori 1979, 258). This conflation of lover
and guest, and the acceptance of the ability of a lover-guest to hurt, again gives
the host a higher position. This returns us to Chakrabarty's (2018) explana-
tion of Derrida's proposal: to love is greater than to be loved. Derrida draws
this from a contretemps "structure of friendship," *contre* meaning against
(Chakrabarty reads this as "other") and *temps* meaning time (101). These questions of time, whether fluid or contradictory, reveal the critical moments when
people step out of the "security and comfort of old affiliations and identifica-
tions" and, in an "unexpected 'gesture' of friendship," make themselves vulnerable to "others" (Gandhi 2005, 188).

Toward Love: An Anticolonial Universalism of the Oppressed

There is a third story to add here, one that brings the two earlier encounters together to work toward an anticolonial politics of climate justice. To bring humanity together in the face of a climate crisis, we could draw on these instances of love, friendship, and hospitality, to build, following Fanon, a new universalism, one that belongs to the oppressed—a universalism that will not deny any particularism that enriches it, and therefore, rather than colonizing, a universalism that is liberating/decolonizing (Chakrabarty 2018, 156).

I return to Leela Gandhi (2014) for a story of South Asian soldier-travelers who fought for the colonizing West against fascism in Europe. In 1914, when Britain joined World War I, most Indian nationalists, led by Mohandas Gandhi and Muhammad Ali Jinnah, halted the anticolonial movement and joined the war effort. Strategically this presented the Indian nationalists with a chance for an "equal" collaboration with Britain. More importantly, this was a moment to rescue "the endangered spirit of Europe from the clutches of imperial-fascism," even if it came "at the cost of immediate, anticolonial nationalist aims" (Gandhi 2014, 96–97). The colonial conscripts drawn from India, West Africa, Algeria, Morocco, Australia, South Africa, and many other countries, even though oppressed by a European universalism that degraded anyone not European, fought in Europe to save the spirit of universalism. This "inter-civilisational ethics" might open a path for an anticolonial universalism, a universalism of the oppressed (Gandhi 2014, 97). This is the universalism that might help us forge an epochal consciousness for climate justice.

Most peasant conscripts, untouched by an elite nationalist rhetoric, joined World War I to earn an income, but with new experiences of conviviality and hospitality in colonial Europe, ended up developing their own understandings of solidarities. What makes this instance of the Great War similar to our current instance of the climate crisis is that many of these subaltern conscripts saw the specter of war in Europe as an end of the world (Gandhi 2014, 98)—an imaginary that no doubt solidified their solidarities for a "common cause."

The peasant conscripts recount not only the food and shelter extended to them by European families but also attempts made by the hosts to accommodate cultural sensibilities, foods, and habits. This experience of hospitality was alongside their experiences of Europe as a place of plenty—exemplified by fields flowing with crops, trees full of fruits, and barns packed with harvest. Gandhi (2014), then, explains the outstanding European generosity "as the moral effect of plenitude: a virtue appropriate to affluence" (99). Whether in the form of hospitality of the local population or care received in military hospitals, for these soldiers, "'white' generosity manifests itself under the sign of excess"—"care achieves the status of a gift on the strength of a humanizing superfluity" (99). On the other hand, sacrifice, for the soldiers, emerged from the conditions of scarcity. Most of these peasant conscripts came from poorer families who needed money to keep their lives and livelihoods going.

For Gandhi (2014), the "sense of collaboration between coloniser and colonised against a common crisis" came from an understanding that "the protection and defence of an admittedly partisan Europe was instrumental for the salvation of the wider world" and was deployed through "the complementary virtues of generosity and sacrifice" that emerged from the "demographically discrete backgrounds of plenitude and scarcity" (101). This instance demonstrates that an anticolonial "moral partnership between congenial antagonists" is possible, even if based on "contradictory etymologies" (101). The colonizer commits to hospitality on the back of "an apologetic idiolect of regard and relationality," and the colonized takes the chance to "speak a language of disregard and invulnerability" (101).

Reflecting on this unusual but effective alliance, Gandhi (2014) asks if we might understand this as people from very different cultures and backgrounds responding "alike to their own victimisation by Europe by seeking to reform Europe" (103).[5] Taking a cue from Gandhi and reflecting on the climate question at hand, one might ask: Is it possible to develop a planetary ethics of climate justice that takes "heart precisely from individual, groups, and cultures" that globalization—the homocentric analogue of planetary—has attempted to dominate and decimate?[6]

I propose that this might be where we initiate a politics of epochal consciousness—one that speaks to humans as a geological force, of one humanity. Here both mitigation of and adaptation to the climate crisis depends on shedding old affiliations and opening ourselves to unexpected friendship and hospitality toward others. These ideas, derived from some radical moments of subaltern politics, seek to open the path for a universalism of the oppressed—a universalism that can help build an epochal consciousness and zoecentric politics. Following Kathryn Yusoff (2018), when we think of the Anthropocene man, we cannot leave out the colonial man. The legacy of the Anthropocene is deeply infused with the legacy of colonization. To build a genuine solidarity that situates justice at the center of the climate crisis, the Anthropocene man needs to be thought of from the vantage of the colonized man or the inhuman/ the lesser-than-human. To develop a zoecentric politics, that is, a life-centric politics, the first step, then, is equity for those who have been degraded and discriminated against as lesser human life forms, lives that colonial and contemporary evidence tells us have mattered less—groups whose behavior and characteristics are racialized and ridiculed as being closer to nature, closer to undisciplined life.

A politics of philoxenia opens us to new ways of thinking about coping with, adapting to, and mitigating climate change. To rethink our response to the climate crisis in a postcolonial space, we need to think through new ideas like hospitality, friendship, and love, as we see in the case of the Dutch electric car traveler. If it were not for these ideas, their journey to prove the possibility of a low-carbon world would not have been successful. Hospitality and friendship helped them prove that we can mitigate climate change. We must remain open to the same hospitality and friendship to adapt to a climate-changed world when the crisis might force even more people to look for new places to survive. I am thinking of guests who might arrive at our doorsteps in a climate-changed urban world (and we already see this happening).

Might we be able to build a community that operates within but without states, one that is open to guests, friendship, and hospitality, to cope with, and adapt to a climate-changed world? If urban areas around the world can take initiatives independent of states to tackle the climate crisis by thinking of ways to contribute to "global" mitigation and "local" adaptation, might they also be able to think of "global" adaptation? After all, the climate crisis is a shared crisis, on all levels.

Conclusion: A City without a State?

When it comes to emission of greenhouse gases, cities can contribute to global mitigation, independent of states. This is because the earth's atmosphere is an open, shared space where the flow of carbon emissions, and the warming they cause, do not respect political boundaries. In cases where people might be forced to migrate, climate adaptation would be premised on sharing resources and space. This is hindered by political borders of states that are maintained by surveillance and carceral infrastructures. The sanctuary cities movement is often credited as an example of refuge that cities can extend within, but without, states. Sanctuary cities offer support and "safe space," often for undocumented migrants, refugees, and those fleeing violence, by "providing identification, housing, health and labour market [and] . . . non-collaboration with deportation agencies" (Fischer and Jørgensen 2020, 5). Yet Ananya Roy (2019, 768), mediated by Derrida and Paul Gilroy, links the proponents of sanctuary jurisdictions and those of "expansion of police power." Sanctuary cities, after all, are not premised on a resistance to the state; rather, as Roy explains, they are expressions of resistance to particular parts of the state. For example, in case of the United States in recent years, sanctuary city proponents have used these as a "resistance to Trumpism" (768). Sanctuary cities are premised on the "liberalism" of local governments, including the local police. However, the same police have a long history of brutality as part of "state violence." This history of violence and carceral infrastructure, which often targets mainly Black, Brown, Indigenous, and Dalit bodies, supported by the local state, is starkly at odds with movements for social justice. It is particularly at odds for this chapter's argument centered on a life-centric politics, which is premised on equity for those humans who have been degraded and discriminated against as lesser human life forms. Sanctuary jurisdictions are premised on the same police that have unleashed violence on Black, Dalit, and Adivasi lives.

We are then left with the question: can a community that operates within but without states, one that is open to guests, friendship, and hospitality, flourish in a city? With the intimacy of "state protection and state violence," Roy (2019) calls for a "shift from sanctuary to abolition" (775). A politics of philoxenia, premised on making the self vulnerable to "others" (Gandhi 2005, 188), is a politics that rejects all manners of "security." A politics without the state is not a politics of sanctuary; it is a politics of abolition.

Postscript

In the earlier sections I have tried to work toward an anticolonial politics of climate justice, one that rejects individualism and builds a community of singularity premised on friendship, hospitality, and love. Yet questions of race, gender, and caste of the colonized abound. Hospitality and love are not easy to come by. Our sociocultural identities get in the way. Here, I leave the readers with one last story of a journey. "Atithi Devo Bhava" (Guest is god) is a story by Indian writer Abdul Bismillah (Sawhney 2018). It recounts the story of a Mohammad Salman, a Muslim man, and a scholar of Sanskrit, who is unable to find a job as a Sanskrit teacher at any institution in postcolonial India.[7] He ends up teaching Sanskrit privately to the son of a local wealthy family. The student moves to the city and invites his teacher for a visit. The teacher makes his visit a surprise, only to find his student's house closed. A neighboring Brahmin family invites him into their house—a guest of the neighbor must be offered hospitality—and even offers food. However, as Mr. Salman is eating, they discover his identity as a Muslim. What was a friendly environment now becomes stressful. The Brahmin family do not throw Mr. Salman out or withdraw their hospitality. However, the metal tumbler in which he was offered water was swiftly changed with a glass tumbler.[8] This is significant. Sawhney (2018) explains that "the hospitality earlier offered is now fearfully withdrawn, the other is marked as other" (2). Derrida (2005) reminds us of the suspension and limits of hospitality, which is a reminder of an opening but also a closure "to protect against the unlimited arrival of the other" (6; see also Kumar 2015). Analyzing Bismillah's "Atithi Devo Bhava," and drawing on Derrida, Sawhney argues that "discrimination, which is on the one hand necessary for hospitality, is also at the same time that against which hospitality must constantly struggle, in order to remain true to its innermost essence" (16).

The hospitality, love, and friendship recounted in my story of the Dutch traveler are similarly racialized. He is a White European. One might recall what happens when a Bangladeshi or a Black African migrates to India. In fact, the arguments above need to be put in perspective of the discrimination many Indians who are racialized as not "pure enough" or "civilized enough" or "Indian enough" face in parts of India.

NOTES

1. Guide to the Harriet Brainard Moody Papers, 1899–1932, https://www.lib.uchicago.edu/e/scrc/findingaids/view.php?eadid=ICU.SPCL.MOODYHB.

2. This is another opening for the zoecentric politics that the Anthropocene demands.

3. She was in awe of his intelligence and therefore was open to his eccentricities.

4. Born partly out of my own cultural and linguistic limitations.

5. Victims of colonialism on the one hand and fascism on the other.

6. Gandhi's original question on colonization: Could we fabricate a Europe that took heart precisely from the individuals, groups, and cultures it had tried so hard to extricate from its civilizational fabric?

7. Although this is a story from the 1980s, similar cases still occur.

8. Many higher-caste Hindu families keep separate sets of utensils to serve those they may consider "impure" according to the Hindu belief system. These utensils are often glass or china as they have lower purity status than metal. This "impure" characterization imparted by higher-caste Hindus mainly applies to Dalits, Muslims, and Christians in India.

REFERENCES

Chakrabarty, D. (2012a). Postcolonial studies and the challenge of climate change. *New Literary History*, 43(1), 1–18.

———. (2012b). Radical histories and question of enlightenment in the writing of modern Indian history. In V. Chaturvedi (Ed.), *Mapping Subaltern Studies and the Postcolonial* (pp. 256–280). Verso.

———. (2014). Climate and capital: On conjoined histories. *Critical Inquiry*, 41(1), 1–23.

———. (2015). The human condition in the Anthropocene. *Tanner Lectures in Human Values*, 139–188. doi: 10.1016/j.quaint.2014.11.045.

———. (2018). *The Crisis of Civilisation: Exploring Global and Planetary Histories*. Oxford University Press.

Derrida, J. (2005). The principle of hospitality. *Parallax*, 11(1), 6–9.

Fischer, L., and Jørgensen, M. B. (2020). "We are here to stay" vs. "Europe's best hotel": Hamburg and Athens as geographies of solidarity. *Antipode*, 53(4), 1062–1082. doi: 10.1111/anti.12707.

Gandhi, L. (2005). *Affective Communities: Anticolonial Thought and the Politics of Friendship*. Duke University Press.

———. (2014). *The Common Cause: Postcolonial Ethics and the Practice of Democracy, 1900 to 1955*. University of Chicago Press.

Kumar, A. (2015). Cultures of lights. *Geoforum*, 65, 59–68. doi: 10.1016/j.geoforum.2015.07.012.

Nandy, A. (2009). *The Intimate Enemy: Loss and Recovery of Self under Colonialism*. 2nd ed. Oxford University Press.

Pachori, S. S. (1979). The Chāyāvād of Java Śankar Prasad: An overview and evaluation. *Journal of South Asian Literature*, 14(1), 247–262.

Roy, A. (2019). The city in the age of Trumpism: From sanctuary to abolition. *Environment and Planning D: Society and Space*, 37(5), 761–778.

Rubin, D., and Prasad, J. S. (1979). The Guest (Atithi). Journal of South Asian Literature, 14(1), 103. https://doi.org/http://www.jstor.com/stable/40872244.

Sawhney, S. (2018). Religion and hospitality in the modern: Thinking with Abdul Bismillah. In L. Choukroune and P. Bhandari (Eds.), *Exploring Indian Modernities: Ideas and Practices* (pp. 211–30). Springer.

Yusoff, K. (2018). *A Billion Black Anthropocenes or None*. University of Minnesota Press.

CHAPTER 12

Confronting Privilege

The Radical Potential of Eco-communities for Urban Climate Justice

JENNY PICKERILL

The problematic positionality of eco-communities as at once activist movements on the margins of formal climate governance, creating vibrant alternative futures, but simultaneously, as movements able to secure themselves against climate change by building retreat spaces for their future, remains largely unexplored (Anderson 2017). Eco-communities deliberately employ low-tech, grassroots, low-cost alternatives, yet the outcomes of their attempts at transformation are worryingly similar—eco-enclaves, exclusionary spaces, rising house prices, and so on.

Privilege in eco-communities is often assertively rejected by participants who deny their class or racial privilege, and instead assert their lack of financial assets as a way to erase other aspects of their relative security. There remain staunch silences around environmental racism from eco-communities—silences about inequity, insecurity, and injustice, facilitated by many residents positioning themselves as the precarious. This chapter explores the damage that this positioning of privileged-as-precarious does to the ability of eco-communities to effectively contribute to transformations required in new climate urbanisms. We must acknowledge that all are not equally exposed to climate change in eco-communities, and in so doing identify more clearly what eco-communities do, and could do more of, in contributing to climate security and equity for all in a just city. Eco-communities need to confront and use their own privilege in order to more effectively contribute to the sociomaterial transformation required for climate justice.

This chapter draws on a range of empirical examples, predominantly from the United Kingdom, United States, and Australia. There is an obvious focus here on Anglophone countries. In addition, as an English white woman, I share much of the privilege I am seeking to challenge. It should be the work of the privileged to do the work of social change in eco-communities, to listen to critiques about exclusionary practices, and not to expect marginalized others to do additional work to make these spaces inclusive. Therefore, this chapter is purposefully focused on the urgent need to enact urban climate justice in the Global North, starting from the position of challenging the very centers of privilege itself.

I begin by exploring the ways eco-communities are progressive spaces of climate mitigation and adaptation. In this, eco-communities can be thought of as spaces of resistance, particularly to neoliberal values and practices, and societal norms, where alternatives are articulated and embedded in the everyday in seeking to mitigate climate change. Crucially, although there are strong linkages between environmental activism in eco-communities, this is not a given. Indeed, many residents in those communities I have worked with do not self-identify as activists nor have activist pasts. Their activism can be considered as somewhat "quiet" or "implicit" (Horton and Kraftl 2009). This is important because, as will be explored in this chapter, although eco-communities have potential to create futures otherwise, many also have a narrow inward-looking focus that does not necessarily center questions or priorities in the same way that intersectional climate justice activism might. In other words, while eco-communities are spaces of hope for urban climate justice, they are rarely primarily concerned with enacting climate justice. They are spaces of resistance, but not necessarily in the same ways that environmental activism is per se. This is the tension that this chapter explores—that eco-communities do offer potential to tackle climate injustice, but they can also be spaces of exclusion and privilege. This privilege is rarely acknowledged, precisely because social justice is not a core quest for many eco-communities. While environmentalism is a key aim of eco-communities, resisting forms of privilege and elitism is too often separated from this political intent. Therefore, eco-communities need to be explicitly anti-racist if they are to contribute effectively to new climate urbanisms that adequately generate socioecological transformations. Otherwise, there is a real risk that their form of environmental protection is little more than the elite looking after themselves. Intersectional social justice has to be central to any environmental sustainability goals.

There are a variety of practices of exclusion at play in eco-communities through which they become rather homogenous, lacking racial or class diversity in particular. These processes interweave and overlap and will be demonstrated here through brief analysis of the (lack of) diversity of bodies, races, and wealth, tied together by an acknowledgment of the ways in which eco-communities can replicate neoliberal rationalities. Exploring why this is the case and the failure to tackle these exclusions, followed by a reflection on what can be done about it, is the focus of this chapter.

Eco-communities as Progressive Climate Urbanism

Eco-communities are collective responses to climate change. They are spaces where alternative types of homes, livelihoods, transport, education, and food provision are experimented with in an effort to both mitigate environmental impact and adapt to climate-changed futures. Eco-communities are already-existing and in-place examples of what is possible in creating a just city. They are grassroots, community-led, holistic, and interdependent (in the issues they tackle and how they tackle them) and are often grounded in an ethics of inclusivity, sharing, and commons (Sanford 2019). Although more prevalent in rural spaces (where land costs are often lower and in some places planning regulations more permissive), there are an increasing number of eco-communities in cities (Ergas 2010; Daly 2017). Examples include Los Angeles Eco-Village (LAEV), Kailash Eco-Village, and Columbia Eco-village in the United States; Christie Walk and Cascade Co-housing in Australia; (see figure 12.1) and Springhill Co-housing in the United Kingdom (Pickerill 2016; Cooper and Baer 2019).

Eco-communities are defined as ecological and community-oriented projects that seek to balance human needs with environmental protection, and can therefore include eco-villages, intentional communities, cohousing, and low-impact developments. Common to these examples is an emphasis on self-built physical and social infrastructures that provide for residents' basic needs using low-tech, low-budget, low-skill systems, often partially off-grid. A core aim for most is to reduce environmental impact by minimizing resource use (in materials, in embodied energy, energy requirements, water use), minimizing the generation of waste (in materials, space, energy, leakage), maximizing the use of renewable energy (such as solar, wind, water), and maximizing use of renewable materials (such as straw, sheep's wool, wood, earth).

New sociomaterialities are formed and enacted through eco-communities. These are dynamic spaces where few things are fixed; instead they are actively making, remaking, unmaking, inventing, and innovating new physical infrastructures (energy provision, water supplies, sewerage, etc.) alongside experimenting in new social systems (Fois 2019; Chatterton and Pusey 2020). It is the collectivity of these projects—the collaborative and self-organized aspects,

FIGURE 12.1. Christie Walk, Adelaide, Australia. Photograph by author

the huge variety of aspects of residents' lives that they share, from finances, to childcare, food production and preparation, education, laundry, maintenance, and the focus on innovative decision-making systems (be that consensus, sociocracy, or others)—that is interesting in exploring how we can build urban climate justice.

It is this emphasis on collectivity that offers the most potential in forging new approaches to tackling climate injustice. Other attempts to create climate "resilient" cities have been driven by top-down technocratic and financially driven investment approaches, which all too often emerge as forms of eco-gentrification (Long and Rice 2019). These spaces often exclude existing residents, increase rental and house prices, and push out informal livelihoods. The focus on technology and changing physical infrastructures is also too often done at the exclusion of thinking through the necessity of new social systems in generating new practices of climate adaptation and in making visible the processes through which climate change is intensified. In other words, while eco-communities can offer some novel and experimental low-tech, low-budget, low-skill systems that could be more broadly adopted across cities, it is really the ways in which they operate collectively (the social formations rather than material interventions) that hold the most promise for urban change. The ways in which collective decision-making, making collective choices for the greater good, developing more inclusive decision-making structures, community ownership of spaces, and collective problem solving are put into practice are vital in urban climate justice efforts.

Eco-communities are often radical interventions into a place that seek to actively resist societal and political norms. They reject other approaches (such as ecological modernism, formal politics, or primarily technological solutions) as being inadequate responses to the threats of climate change, and instead adopt more challenging changes. While often radical in their intent, however, many eco-communities also very deliberately seek to start from what is already in place, the grass roots—buildings, communities, and infrastructures—and work with what is preexisting in reworking what homes and livelihoods could look like. The urban eco-communities that already exist were formed by residents already living locally. Eco-communities therefore start at a different scale, often small and incrementally, slowly building momentum and attracting others interested until enough people have joined to make it a viable project. In the early stages most hold numerous open forums for existing local residents, inviting participation. This approach achieves several things: it opens dialogues in place about why eco-communities are necessary, making visible how existing infrastructures (electricity sources, water use, waste, etc.) and daily practices hasten climate change, and raising questions about our accountability in carbon emissions. Eco-communities often start with very broad intentions and reach, seeking to bring in others into these conversations with an educational ambition. This approach also enables existing residents to have input into what any emerging eco-community will look like, creating relations early on, which it is hoped will generate long-lasting engagement. In practice eco-communities seek to build diverse communities through trying to generate affordable housing, designing spaces that facilitate frequent social interaction and repurposing marginalized public urban spaces for food production. The intent is to make permanent interventions to build and reshape the urban, lasting material, social and economic transformations as a form of insurgent urbanism (Pickerill 2020).

Eco-communities as Spaces of Exclusion

Eco-communities often celebrate their diversity. Despite this noble intent and often positive beginnings, many eco-communities develop a disjuncture between their imagined projects and their realization. While some (Los Angeles Eco-Village; LILAC, Leeds) are well positioned and structured to attract diverse residents, many (EcoVillage at Ithaca, New York; Findhorn, Scotland; Hockerton, England; Lancaster, England, to name just a few) struggle to reach beyond the white upper-middle-class cohort. In the Global North eco-communities are too often dominated by a narrow demographic—often highly educated, white, able bodied, and with a predominance of women (Chitewere 2018; Bhakta and Pickerill 2016). Even when emerging in the Global South there are similar processes of exclusion that delimit them to the wealthier middle classes, certain ethnicities, and often expatriate communities (Silver 2018). Consequently, the outcomes of eco-communities' attempts at transformation can look worryingly similar to other forms of gentrification—eco-enclaves, rising property prices, and exclusionary bounded places—which entrenches rather than ameliorates existing inequalities in similar ways to other forms of climate urbanism (Rice et al. 2020). Chitewere (2018) classifies the EcoVillage at Ithaca as a green gated community, an exclusive commodified space of experiences, and a form of green flight. With any radical project it is therefore vital to ask, "Who or what is really being transformed, and to what ends?" (Last 2012, 710).

As is demonstrated in this discussion of exclusions based on diverse bodies, race, and wealth, eco-communities can replicate neoliberal rationalities. This mirrors other seemingly radical experimental spaces—what might initially appear as alternative forms of transformation can be built on neoliberal rationalities, reproducing neoliberal conditions that undermine their radical potential (Argüelles, Anguelovski, and Dinnie 2017). Indeed, many eco-communities replicate, repeat, and mirror conventional society in multiple ways (gender relations, the way money is used, etc.), and rely on state support. Eco-communities have elements that embrace conventional neoliberal values. At the same time, projects that start with racial difference as a key defining factor have developed more radical alternatives (Bledsoe, McCreary, and Wright 2019).

There is a strong class element to this, but also a presumption of individual empowerment (rather than structured privilege) in being able to reject state infrastructures and welfare. Argüelles, Anguelovski, and Dinnie (2017) summarize such rationalities as a focus on individual responsibility rather than calling for state intervention, which in turn "might help to legitimize neoliberal attempts of disposing the state from its economic and societal functions" (38). This ability to retreat from the state is reliant on the "privileged progressive whiteness that permeate" (40) these experiments, an environmental and social privilege that enables such individuals to self-provide, self-organize, and improve their quality of life.

There is an expectation in most eco-communities that residents need to be physically strong, fit, able bodied, and dexterous. While rationalized in rural eco-communities like Tinkers Bubble (Somerset, UK) as a result of refusing to use fossil fuels and therefore fossil-fueled machinery, assumptions remain in many urban eco-communities about the necessity to be physically able to participate in gardening, food production, building, moving about the site without using vehicles, shared cooking responsibilities, and site maintenance. Requirements are made of participants in terms of time and energy to actively participate in communities and work teams. The "work" in eco-communities, despite their apparently progressive politics, also often remains stubbornly gendered (Pickerill 2015). As Amita Bhakta and I previously explored in an eco-community in England, "Little consideration has been given to those unable to undertake these tasks and the material infrastructure (such as doorways, gates and latches) have all been designed and built to suit a conventional body" (Bhakta and Pickerill 2016, 415). A disabled body therefore struggles with eco-communities' assumptions about particular bodily capacities and issues of dexterity and mobility. Even if buildings are designed to be accessible, rarely is the broader space of the community (i.e., the gardens, the meeting spaces, etc.) so that they become far from the spaces of diverse interaction that they had anticipated. There is continued failure to build eco-communities for diverse and disabled bodies.

Just as disabled bodies are not accommodated in many eco-communities, so too race is often ignored as an issue. In the Global North eco-communities are predominantly white, and the absence of Black participants is rarely critically interrogated. Indeed, it is in some places consciously created, as one interviewee at EcoVillage at Ithaca noted, "You have a lot more . . . control about who your neighbours are" (Chitewere 2018, 95), suggesting that some eco-communities are deliberately created as places of escape from differentiated others. Although there has been little explicit research on racism in eco-communities, they mirror the important analysis of alternative food networks (AFN), another space of radical sociomaterial transformation, as being spaces of white privilege (Lockie 2013). This is not to suggest a lack of Black farmers, Black food networks, or Black social justice food campaigning, but to acknowledge that such spaces are rarely visible in discussions of radical transformatory projects in the same ways that white-dominated AFN's are (Alkon 2012). Even when the inequitable implications of demand for, for example, organic food are proved to rely on precarious work regimes that have racialized inequality built into them, AFN still uses "a moral economy framing [which] can obscure systemic inequities in precarious farm employment and dampen the impetus for structural change through collective food movement organizing" (Weiler, Otero, and Wittman 2016, 1140). In other words, it is dismissed as an unfortunate case of a few "bad apple" farmers, rather than a structural problem.

The inclusion of race is too often tokenistic or through forms of racial-cultural appropriation (such as use of Indigenous spiritual symbols or practices). The lack of racial diversity is explained as an individual failure either for Black people to "want to join" eco-communities or a lack of affordable housing (thereby assuming all Black people have less wealth). The lack of racial diversity is rarely articulated as a complex sociocultural question where structures of belonging, identity, racism, and social justice hinder broader participation, and that eco-communities rarely attempt to tackle racial capitalism (Chitewere 2018). Yet as Joe, a long-term member of Hart's Mill Ecovillage North Carolina argues, often the guise of being radically alternative obscures the realization for participants that they are reproducing white supremacy culture:

> I see many intentional communities reproducing a lot of elements of white supremacy culture, especially the sense of urgency, paternalism, fear of conflict, worship of the written word, and quantity over quality. Consensus and sociocracy are often used in a way that unconsciously reproduces white supremacy culture (and patriarchal culture). What's worse, when participants perceive themselves as doing something different but don't recognize how they are continuing patterns of racial exclusion and dominance . . . so there can be a tension or paradox in groups that are creating something new while reproducing elements of mainstream, patriarchal, white supremacy culture. (Quoted in Cole, Horton, and Pini 2019, 55)

Inequities in wealth are recognized in eco-communities to some extent but tend to only be approached as a problem of affordability of housing at the joining stage of community formation. Several urban eco-communities have explicitly sought to radically reduce the cost of housing thereby challenging the exclusivity of the neoliberal housing market. Yet few have done this in a way that fundamentally challenges the market-based approach to housing in the long term. The majority have started as low cost but failed to prevent a reversion to market valuations of property prices, which has obviously curtailed who can then buy into the community. Others are privately owned properties (Kailash Eco-Village) then rented, or community owned but still rented (Los Angeles Eco-Village [LAEV], Christie Walk), albeit at below market rates. It is only Low Impact Living Affordable Community (LILAC) where all residents pay 35 percent of their income for housing (purchasing shares they can eventually sell) that has sought to prevent long-term cost inflation while also giving residents security and capital growth in their investment (Chatterton 2013).

Eco-communities can reduce the costs of everyday living in important other ways, benefiting from economies of scale in energy-generating infrastructures, having smaller home units but access to large shared spaces and facilities such as laundry rooms, bike storage, cars, gardens, visitors' rooms, and entertaining space (Columbia Eco-village) as well as tools and equipment (Jarvis 2019). Residents can also benefit from shared social responsibilities where LILAC, LAEV, Kailash, Springhill Co-housing (UK), Cascade Co-housing (Australia), and Christie Walk share childcare and elder care, shopping errands, and cooking.

Chitewere's (2018) analysis of EcoVillage at Ithaca (EVI) detailed how the class and identity of its residents acts to exclude differentiated others. Only those "with both the economic and social capital" (140) who have money but only work part-time or from home can join EVI. Seemingly subtle assumptions about participation, common values, lifestyle, and food choices shape who gets to be part of eco-community experiments. Indeed, the very intent to generate collectivity, a core necessity in creating and maintaining eco-communities, can act as a driver of racial division in that the very claim of being equal is only possible in a white space, yet this remains unacknowledged. In this sense, the seeking of collectivity can erode the question of race. Likewise, there is potentially an epistemological distinction that codes the definition of eco-communities as intrinsically white, and in so doing frames other spaces of resistance as not environmental "enough" (see Safransky 2017). In these ways even seemingly politically progressive experimentations can be built on troubling exclusions.

Understanding Privilege

Privilege is a structural advantage that benefits those of particular race, class, gender, or identity categories (Bhopal 2018). Privilege is systematically produced through ongoing processes of dominance and uneven geographies that materially and socially benefit certain populations. Privilege requires attending not just to historical conditions but also the ongoing processes and logics (such as settler colonialism) that continue to support and ensure privilege and the consequent erasure of others. For example, whiteness remains an invisible normative category (Bonnett 2000; Pulido 2015; Joshi, McCutcheon, and Sweet 2015), one that intersects with the logics of class but ultimately "takes precedence over all other forms of identity"; in other words, "the identity of whiteness is . . . the first determinant of how groups are positioned" (Bhopal 2018, 27).

The limitation of privilege as a concept is that it can be a way to claim an innocence (Leonardo 2004; Wekker 2016), to shift responsibility from the individual to the category and simultaneously suggest that its acknowledgment resolves its advantage (Ahmed 2004), without materially changing any existing structures of oppression and death (Gilmore 2002). The term is therefore used here cautiously to articulate a problem, rather than tightly define its remit. As Joe (quoted above) articulates, it is necessary at times to employ more explicit language—such as white supremacy—when reflecting on certain forms of racial privilege and "the presumed superiority of white racial identities" (Bonds and Inwood 2016, 719) that actively produces white privilege and the power relations and material conditions of advantage (Berg 2011). Likewise, authors such as Roediger (2019) have suggested a shift to using the term "white advantage" in an attempt to acknowledge the complexity of how poverty and misery is not always racially discriminating. Rothman and Fields (2020) likewise raise the dilemma that to use white privilege as a broad-brush term potentially alienates those who might otherwise seek productive alliances and commonality with Black fellow citizens. Therefore, the term "privilege" is purposefully used here as a useful tool to start and frame important conversations in eco-communities about structural advantage, while simultaneously acknowledging the limitations of its definition.

Eco-communities are spaces of exclusion in large part because of the ways in which they have failed to acknowledge and respond to their privileges. These privileges are inherited from and structured by some of the broader mainstream environmental activisms that many have emerged from (Taylor 2016). Mainstream environmentalism, for example, is built on a troubling colonial history and a "fortress" conservation approach where nature is protected by the exclusion of humans (Paperson 2014). This can be seen in how some environmental campaigns purposefully focus on saving iconic species (whales, old growth trees, etc.), without necessarily paying enough attention to the social justice implications (people's jobs, traditional practices) of how this will likely disadvantage particular (often racialized) groups of people. In other words, the privilege of whiteness has facilitated the production of the "white savior" environmentalist who determines how others should live, sometimes drawing on Indigenous or Black environmental approaches, ultimately creating stereotypes or co-opting them rather than including them as equals. Consequently, mainstream environmentalism exemplifies a variety of types of "othering" in their discourses and exclusion in their political narratives (Erickson 2020; Paperson 2014), generating a long-held and ongoing suspicion among, for example, Indigenous activists, that the protection of the environment will ultimately be prioritized over Black bodies (Pickerill 2009). This is reflected in eco-communities when their ecological rationale erodes the social justice politics that many of them began with.

The structural advantages that many residents in eco-communities (particularly in the Global North, but globally too) benefit from, particularly through whiteness and class, position participants as relatively wealthy. Even if they have not already secured financial savings (with which to purchase land or homes) they are able to access loans or secure professions (which often also support part-time or working from home), and have access to higher education systems and qualifications. By seeking like-minded participants in order to more easily build common intentionality and collectivity, there is also an ongoing tension between celebrating diversity and yet purposefully seeking homogeneity in order to more easily build community (Christian 2003). At EVI a resident argued that low-income families would complicate communal decision-making because they might hold different values (Chitewere 2018).

Although many of these privileges are acknowledged, they are rarely considered a result of structural advantage but one of individual good fortune (and therefore exclusion is also not a structural problem), and/or are articulated as interchangeable and equivalent with other types of disadvantage that residents experience. The individualization of privilege enables residents to deny that they have benefited from a system such as racial capitalism, that has by definition caused the oppression of differentiated others. It enables participants to reject any obligations or responsibilities for their privilege, a denial of its social injustice implications. As such, as Chitewere (2018) argues, living at EVI appears to enable residents to avoid facing their own contradictions, with many asserting by virtue of membership that they were doing their bit for the greater environmental good. This positioning negates residents having to engage in any acts of real sacrifice or systematic change—the consumption of the eco-village is used to signal that they have contributed "enough" toward socio-environmental transformations, whereas actually they have merely engaged in potentially ineffective forms of performative eco-consciousness.

Taken further, this individualization is also applied by members in explaining a lack of racial diversity in eco-communities. The lack of Black eco-community residents is justified as a lack of environmental concern by predominantly white participants. This is a deeply flawed but often repeated belief that the lack of Black participants in environmental projects and eco-communities is because of a lack of prosperity—that it is only once people have their basic needs met that they can afford to be concerned about the environment (Hickcox 2018; Gomez 2020). There is, of course, plenty of evidence of Black concern for environmental issues, such as Black-led environmental movements and environmental justice organizations (Carter 2016). Indeed, not only have anthologies documented "black nature" (Dungy 2009), but Carolyn Finney (2014) has carefully examined how the concept of outdoor nature and the environment have been racialized and are being reclaimed by African Americans, despite anti-Blackness persisting (Finney 2020).

The reticence of Black involvement in eco-communities is more likely reflective of the unacknowledged white privilege on the part of many residents, a colonial history of conservation and the failure of contemporary environmentalism to acknowledge structural problems, inequality, and the need for social justice. As Chitewere (2018) argues, unless capitalism is structurally challenged, and therefore the broader structures of inequality and exclusion tackled in eco-communities, then issues of diversity, especially of race and class, will not be resolved. In other words, personal or individual environmentalism will not alter the culture of capitalism or result in broad-scale environmental justice. Yet silences remain around environmental racism, inequity, and injustice in eco-communities. Instead, onus is placed on the individual resident to (voluntarily) engage in training, notably not seen as necessary until Black people arrive: "We're encouraging all members to do substantive training in white supremacy and racial equity, but for now it's optional. It remains to be seen what adjustments we may need to make in our governance practices should we start attracting Black members. But we definitely need to raise our awareness of the elements of white privilege and power now" (Hope quoted in Cole, Horton, and Pini 2019, 56).

Furthermore, when privilege is explicitly discussed, eco-community residents have been quick to articulate their insecurity, most often financially. They position themselves as the precarious as a way to reject any privilege. However, such limited financial capital or income is often purposeful—a form of voluntary simplicity or voluntary poverty (Vannini and Taggart 2013) and manufactured vulnerability (Doherty 1999). Even if it is not, such vulnerability is not interchangeable and equivalent with that of white supremacy (as argued, because race dominates as a privilege), and it cannot be allowed to erase residents' multiple other forms of privilege.

Conclusion: Challenging the Privileged-as-Precarious

Eco-community residents are quick to individualize and seek to reduce notions of privilege, thereby erasing their structural advantages. This perpetuates racial exclusion and class exclusivity, and other key aspects of any necessarily intersectional environmentalism: "There are limits on how cooperative a group can be if it's bringing old habits and practices from the dominant culture of competition. There are always many dimensions of power, rank and privilege present in any human group, and it's important to cultivate individual and group consciousness about these power differences, and have conversations about how to address them.... Both [decision-making systems—consensus and sociocracy] may therefore entice participants into a false sense of 'instant equality' without addressing the unequal power relations within the group" (Joe quoted in Cole, Horton, and Pini 2019, 53).

An absence of a critical analysis of privilege and power in eco-communities means the broader political possibilities of transformative change are limited. There is a need to be vigilant to the politics of eco-communities and to their justice. Eco-communities offer significant potential as forms of progressive climate urbanism, but only if the structural dynamics of privilege are acknowledged and challenged. Otherwise, eco-communities can become time and energy sinks that distract from participating in broader social justice and environmental struggles and instead become focused on protecting the privileged few. This, of course, limits eco-communities' capacity for resistance to neoliberalism and social and political norms, or rather forecloses which elements of contemporary society are challenged and which are left as is. This produces new urban spaces that might be more ecological but also more privileged.

White-centered eco-communities should use their privilege more productively to collectively challenge conventional ways of being and organizing, and more assertively reject the "old habits and practices" that Joe laments (Cole, Horton, and Pini 2019). This is not only about "letting more Black people in," or making homes "more affordable," or even ensuring that decision-making forums or work distribution processes are more inclusive. It requires eco-communities to become assertively anti-racist (Kendi 2019), to use their privilege, resources, and power to share and reach out and join Black-led initiatives. This is a multistaged process that requires the privileged to acknowledge the problem and move past their own discomfort to be able to even see privilege and whiteness. As Rios (2020) argues, in putting into practice Nieto's (2014) strategy, there are many missteps on this route, which often includes inviting "Black, Indigenous, People of Color (BIPOC) to join white-centered spaces without attempting to make these spaces more welcoming." It is only when "we become effectively anti-racist through participating in individual, institutional, and structural change that is envisioned and led by BIPOC" (55) that effective change can begin.

This radical anti-racist praxis should draw on the work of Black radical feminism, which enables the structures of privilege to be questioned (Lorde 2017; Oluo 2019). This involves centering Black people and following their lead, creating Black spaces, running trainings for residents about cultural appropriation, racism, white supremacy, and so on, and exploring what reparations might look like (Johnson and Wilkinson 2020). It requires building new forms of radical relationality.

In practical terms there are numerous starting points for how to do anti-racist work in eco-communities. These connect directly to the possible urban future of "secure and equitable" (Rice, Levenda, and Long, this volume), where the process of resisting climate gentrification, for example, requires focusing efforts on systemic change in landownership, housing, use of public spaces, local livelihoods, and welfare provisions. This requires resisting the material and social processes that facilitate the unevenness of the impacts of climate change and instead generating alternatives.

Building anti-racism requires internal reflection and external action—actively asking Black activists what the priority issues are for them and cocreating equitable solutions to problems. This simple step of asking and then listening to, not assuming, what matters, and then acting on these issues, has been vital in the praxis of anti-racism. Just as environmentalism has to reconsider how it defines an "environmental problem" to incorporate a more diverse set of issues (for example, beyond biodiversity to questions about job creation), so eco-communities have to understand what their neighbors' priority concerns are.

There are many lessons that can be learned from work in environmentalism that has challenged racism, such as in explicitly Black environmental groups including Black2Nature, Black Environmental Network, Black Girls Hike, Outdoor Afro, Green Worker Cooperatives, and the specifically climate-focused group Climate Reframe. Directly reaching out to these groups and offering to share resources (material but also time and energy) in supporting their initiatives is just a first step. Likewise, while "inviting in" to white spaces is not enough, if eco-communities include green spaces, gardens, and natural play areas, then sharing access to these is an important step in acknowledging the inequity of engagement with nature. Likewise, making it clear that communal spaces, meeting rooms, and eco-community resources can be used for initiatives and projects of Black activists' choosing is a simple inclusive step.

A bigger and vital step is to consider what eco-communities generate that could be shared. This could include renewable energies, locally grown organic food, childcare, editing and publishing abilities, or skills and expertise. Or it could mean sharing the residents' privilege to give Black activists media platforms, stepping aside and letting other people speak, cocreating possibilities of funded positions in eco-communities, and validating the lived experiences of Black activists in determining how to enact social and environmental change. This approach requires that eco-communities ask themselves what contributions they can make in greening the city, new environmentally sustainable job creation, or the transition to democratic, collectively owned energy systems.

This, then, is the radical potential of eco-communities for urban climate justice—spaces that not only offer social and environmental alternatives in practice, that generate spaces for experimentation and hope in a climate-changed world, but also confront their privilege through anti-racist praxis and in so doing open up the urban to new possibilities.

REFERENCES

Ahmed, S. (2004). Declarations of whiteness: The non-performativity of anti-racism. *Borderlands*, 3(2).

Alkon, A. H. (2012). *Black, White, and Green: Farmers Markets, Race, and the Green Economy*. University of Georgia Press.

Anderson, J. (2017). Retreat or re-connect: How effective can ecosophical communities be in transforming the mainstream?. *Geografiska Annaler B: Human Geography*, 99(2), 192–206.

Argüelles, L., Anguelovski, I., and Dinnie, E. (2017). Power and privilege in alternative civic practices: Examining imaginaries of change and embedded rationalities in community economies. *Geoforum*, 86, 30–41.

Berg, L. (2011). Geographies of identity I: Geography–(neo)liberalism–white supremacy. *Progress in Human Geography*, 36(4), 508–517.

Bhakta, A., and Pickerill, J. (2016). Making space for disability in eco-housing and eco-communities. *Geographical Journal*, 182(4), 406–417.

Bhopal, K. (2018). *White Privilege: The Myth of a Post-racial Society*. Policy Press.

Bledsoe, A., McCreary, T., and Wright, W. (2019). Theorizing diverse economies in the context of racial capitalism. *Geoforum*, 132, 281–290.

Bonds, A., and Inwood, J. (2016). Beyond white privilege: Geographies of white supremacy and settler colonialism. *Progress in Human Geography*, 40(6), 715–733.

Bonnett, A. (2000). *White Identities: Historical and International Perspective*. Prentice Hall.

Carter, E. D. (2016). Environmental justice 2.0: New Latino environmentalism in Los Angeles. *Local Environment*, 21(1), 3–23.

Chatterton, P. (2013). Towards an agenda for postcarbon cities: Lessons from Lilac, the

UK's first ecological, affordable cohousing community. *International Journal of Urban and Regional Research*, 37(5), 654–1674.

Chatterton, P., and Pusey, A. (2020). Beyond capitalist enclosure, commodification and alienation: Postcapitalist praxis as commons, social production and useful doing. *Progress in Human Geography*, 44(1), 27–48.

Chitewere, T. (2018). *Sustainable Community and Green Lifestyles*. Routledge.

Christian, D. L. (2003). *Creating a Life Together: Practical Tools to Grow Ecovillages and Intentional Communities*. New Society.

Cole, J., Horton, H., and Pini, M. (2019). Culture change or same old society? Consensus, sociocracy, and white supremacy culture. *Communities*, 184, 53–56.

Cooper, L., and Baer, H. A. (2019). *Urban Eco-Communities in Australia: Real Utopian Responses to the Ecological Crisis or Niche Markets?*. Springer.

Daly, M. (2017). *The most powerful form of activism is just the way you live: Grassroots intentional communities and the sustainability of everyday practice*. University of Technology Sydney, PhD thesis.

Doherty, B. (1999). Manufactured vulnerability: Eco-activist tactics in Britain. *Mobilization: An International Quarterly*, 4(1), 75–89.

Dungy, C. T. (Ed.). (2009). *Black Nature: Four Centuries of African American Nature Poetry*. University of Georgia Press.

Ergas, C. (2010). A Model of sustainable living: Collective identity in an urban ecovillage. *Organization and Environment*, 23(1), 32–54.

Erickson, B. (2020). Anthropocene futures: Linking colonialism and environmentalism in an age of crisis. *Society and Space D*, 38(1), 111–128.

Finney, C. (2014). *Black Faces, White Spaces: Reimagining the Relationship of African Americans to the Great Outdoors*. UNC Press.

———. (2020). The perils of being black in public: We are all Christian Cooper and George Floyd. *Guardian*, June 3. https://www.theguardian.com/commentisfree/2020/jun/03/being-black-public-spaces-outdoors-perils-christian-cooper.

Fois, F. (2019). Enacting experimental alternative spaces. *Antipode*, 51(1), 107–128.

Gilmore, R. W. (2002). Fatal couplings of power and difference: Notes on racism and geography. *Professional Geographer*, 54, 15–24.

Gomez, K. L. (2020). Confronting white environmentalism. *Toyon Literary Magazine*, 66(1), 22.

Hickcox, A. (2018). White environmental subjectivity and the politics of belonging. *Social and Cultural Geography*, 19(4), 496–519.

Horton, J., and Kraftl, P. (2009). Small acts, kind words and "not too much fuss": Implicit activisms. *Emotion, Space and Society*, 2(1), 14–23.

Jarvis, H. (2019). Sharing, togetherness and intentional degrowth. *Progress in Human Geography*, 43(2), 256–275.

Johnson, A. E., and Wilkinson, K. K. (Eds.). (2020). *All We Can Save: Truth, Courage, and Solutions for the Climate Crisis*. One World.

Joshi, S., McCutcheon, P., and Sweet, E. L. (2015). Visceral geographies of whiteness and invisible microaggressions. *ACME*, 14(1). https://acme-journal.org/index.php/acme/article/view/1152.

Joubert, K., and Dregger, L. (2015). *Ecovillage: 1001 Ways to Heal the Planet.* Triarchy Press.

Kendi, I. X. (2019). *How to Be an Antiracist.* One World.

Last, A. (2012). Experimental geographies. *Geography Compass*, 6(12), 706–724.

Leonardo, Z. (2004). The color of supremacy: Beyond the discourse of "white privilege." *Educational Philosophy and Theory*, 36, 137–152.

Lockie, S. (2013). Bastions of white privilege? Reflections on the racialization of alternative food networks. *International Journal of Sociology of Agriculture and Food*, 20, 409–418.

Long, J., and Rice, J. L. (2019). From sustainable urbanism to climate urbanism. *Urban Studies*, 56, 992–1008.

Lorde, A. (2017). *Your Silence Will Not Protect You.* Silver Press.

Nieto, L. (2014). *Beyond Inclusion, beyond Empowerment: A Developmental Strategy to Liberate Everyone.* Cuetzpalin.

Oluo, I. (2019). *So You Want to Talk about Race.* Hachette.

Paperson, L. (2014). A ghetto land pedagogy: An antidote for settler environmentalism. *Environmental Education Research*, 20(1), 115–130.

Pickerill, J. (2009). Finding common ground? Spaces of dialogue and the negotiation of Indigenous interests in environmental campaigns in Australia. *Geoforum*, 40(1), 66–79.

———. (2015). Bodies, building and bricks: Women architects and builders in eight eco-communities in Argentina, Britain, Spain, Thailand, and USA. *Gender, Place and Culture*, 22(27), 901–919.

———. (2016). *Eco-homes: People, Place, and Politics.* Zed Books.

———. (2020) Making climate urbanism from the grassroots: Eco-communities, experiments, and divergent temporalities. In V. Castán Broto, E. Robin, and A. While (Eds.), *Climate Urbanism: Towards a Critical Research Agenda.* Palgrave Macmillan.

Pulido, L. (2015). Geographies of white supremacy and ethnicity 1: White supremacy vs. white privilege in environmental racism research. *Progress in Human Geography*, 39(6), 809–817.

Rice, J. L., Cohen, D. A., Long, J., and Jurjevich, J. R. (2020). Contradictions of the climate-friendly city: New perspectives on eco-gentrification and housing justice. *International Journal of Urban and Regional Research*, 44(1), 145–165.

Rios, M. (2020). Centering Blackness in our soils and our souls to promote climate justice. *Communities: Life in Cooperative Culture*, 187, 53–57.

Roediger, D. (2019). White privilege, white advantage, white and human misery. Verso Books blog, March 8. https://www.versobooks.com/blogs/4262-white-privilege-white-advantage-white-and-human-misery.

Rothman, A., and Fields, B. J. (2020). The death of Hannah Fizer. *Dissent*, July 24. https://www.dissentmagazine.org/online_articles/the-death-of-hannah-fizer.

Safransky, S. (2017). Rethinking land struggle in the postindustrial city. *Antipode*, 49(4), 1079–1100.

Sanford, W. A. (2019). *Living Sustainably: What Intentional Communities Can Teach Us about Democracy, Simplicity, and Nonviolence.* University Press of Kentucky.

Silver, J. (2018). Suffocating cities: Urban political ecology and climate change as social-ecological violence. In H. Ernstson and E. Swyngedouw (Eds.), *Urban Political Ecology in the Anthropo-obscene: Interruptions and Possibilities* (pp. 129–146). Routledge.

Taylor, D. (2016). *The Rise of the American Conservation Movement: Power, Privilege, and Environmental Protection.* Duke University Press.

Vannini, P., and Taggart, J. (2013). Voluntary simplicity, involuntary complexities, and the pull of remove: The radical ruralities of off-grid lifestyles. *Environment and Planning A,* 45(2), 295–311.

Weiler, A. M., Otero, G., and Wittman, H. (2016). Rock stars and bad apples: Moral economies of alternative food networks and precarious farm work regimes. *Antipode,* 48(4), 1140–1162.

Wekker, G. (2016). *White Innocence: Paradoxes of Colonialism and Race.* Duke University Press.

CONCLUSION

Toward Transformative
Urban Climate Justice

Abolition, Care, and Reparations

ANTHONY LEVENDA, JENNIFER L. RICE,
AND JOSHUA LONG

The chapters in this book present a range of perspectives on urban climate justice to inform theory, praxis, and resistance. Our shared goal has been to provide ideas and practices that might help us create more equitable, just, and democratic cities in the era of climate change. Some chapters called for radical shifts away from conventional ideas about how knowledge on climate change is produced (Castán Broto et al.) and how urban land and property is governed under a changing climate (Shi and Bouma). Others considered the role of climate finance, funding, and debt in the perpetuation or challenging of status quo climate politics (Silver and Cox). The work that eco-communities might do for racial justice was examined (Pickerill) and the opportunities for new forms of friendship and hospitality to open up anticolonial forms of climate activism was also explored (Kumar). Importantly, several authors demonstrated the importance of putting theory into practice—that is, finding ways to break down traditional barriers between the academy and policy, planning, and activism. Two chapters (Fitzgerald, Schmitz, and Stephens; and Leichenko, Foster, and Nguyen) provided new approaches to evaluating the plans and actions of local governments on climate change to center transformative ideas of justice (in particular, racial justice). Other chapters offered thoughtful and important reflections on doing the actual work of climate activism and advocacy through meaningful collaborations with marginalized groups (Goh and Raditz), including challenging the racial state, racial capitalism, and settler colonialism as a necessary move toward climate justice (Martinez-Lugo). In each case, we see clearly how on-the-ground social action is essential for confronting the climate crisis on a broad scale.

As noted in the introductory chapter, the authors in this volume are primarily situated in U.S. and British universities and knowledge production contexts. This offers most of us the privilege to speak about climate change with scholarly authority, and as a result of our position, the majority of contributors remain largely insulated from the most harmful effects of the climate crisis. This certainly contributes to a partial perspective presented in this volume. Yet we have attempted to deploy various aspects of our privilege in ways that seek to transform underlying institutions of oppression and inequality that contribute to climate change, and also underpin the production of precarity, violence, and harm in other ways. We hope that the diversity of approaches to this normative and political project of urban climate justice facilitates a more in-depth and central engagement with the experiences and expertise of those marginalized and oppressed people and communities on the front lines of climate change. Authors in this volume have done this through a variety of means, including research collaborations that center the needs of vulnerable communities, activities of scholar-activism, engaging theory from the Global South and Black, Brown, Indigenous, queer, and female scholars, creating and using subversive methodologies, and critically evaluating established and prominent institutions of climate governance as "experts" in the field.

If we return to figure o.1 of the introduction, we can see that there are several points of intersection about how to create a more secure and equitable urban climate future that emanate from the work presented here. As such, the task of this concluding chapter is not solely to summarize the range of ideas presented, but rather to synthesize and imagine futures of urban climate justice around key themes from this book. With this task, we outline three areas of continuing work that will push our efforts to not only imagine and theorize but to put these radically new imaginaries of the climate-just city into practice. In these final pages, we focus on the themes of abolition, politics of care, and reparations. Each of these concepts is offered not as abstract thinking but as real practices that can (and must) be implemented to create the just city in the era of climate change. In other words, we explore a series of interventions for climate justice that can be connected to the cities in which we live.

Abolition

Throughout this book, we have argued that climate justice is fundamentally about addressing root causes of injustice and oppression. We show how climate injustices are produced by the same processes that contribute to police brutality, persistent poverty and income inequality, gentrification and displacement, and the devaluation and erasure of marginalized peoples. We and many contributors to this volume take inspiration from traditions of Black radicalism and decolonization to examine the processes of racial injustice that contribute not only to climate change but also to a range of persistent issues in our cities. To-gether these approaches narrate potentialities within, against, and beyond ra-cial capitalism and the settler-colonial state. We chart futures that might aptly be described as "horizons" of urban climate justice that are emergent within long-standing resistances to these oppressive systems. The conception of abo-lition geographies put forth by Ruth Wilson Gilmore (2017) is formative to our thinking here, where she notes that the central task of abolition geography is "to find alternatives to the despairing sense that so much change, in retrospect, seems only to ever have been displacement and redistribution of human sacri-fice. If unfinished liberation is the still-to-be-achieved work of abolition, then at bottom what is to be abolished isn't the past or its present ghost, but rather the processes of hierarchy, dispossession, and exclusion that congeal in and as group-differentiated vulnerability to premature death" (228).

Climate change exacerbates processes leading to group-differentiated vul-nerability and to premature and slow death. We have already witnessed the un-folding of these processes, as many chapters in this volume attest. This has led to calls for abolitionist climate justice that situate current projects for climate resilience, climate adaptation, and climate mitigation within the broader his-torically rooted processes that create expendable populations (Dawson 2010). For example, Ranganathan and Bratman (2021) explain, "Rather than assign blame for vulnerability on individual bodies and deficient behaviors via indi-cators like 'poverty,' 'obesity,' and 'lack of education,' as expert climate plans tend to do, it is necessary to shift the gaze to the historical and multi-causal production of harms" (132). Abolitionist approaches necessarily account for the social production of vulnerability to climate change via interconnected op-pressions including racism, classism, and sexism. Elsewhere, we have argued this requires a rejection of an emerging system of climate apartheid, rooted in the expendability of oppressed groups (Rice et al. 2022). Indispensability of all people, we and the authors of this volume have argued, is central to the prac-tice of abolition, whether it be to reject violent fossil fuel energy systems sup-ported by settler-colonial states, to end environmental racism in climate disas-ter response, or to discontinue the overreliance on technocratic and colonial forms of knowledge to understand and address the problem of climate change.

Of course, diagnosing the origins of climate injustice is only a first step. We need abolitionist praxis. Part of the work of academics and activists is to inte-grate anti-racist, decolonial, and feminist approaches into our climate action. This is already starting to take hold. The climate justice movement has been adamant about calling out the root causes of injustice, even as mainstream climate technocrats flatten various categories of social difference. Recent efforts—especially by youth activists—have been at the forefront of radical climate justice movements that demand a different world. For example, leaders of groups such as the Sunrise Movement and Zero Hour highlight the need to abolish underlying systems of oppression that structure climate injustice. The proliferation of activist movements that call for broad system change resonate with abolitionist struggle in search of climate justice.

Abolition, then, is not a simple nod to the need for something different—it is an ongoing struggle against the deep-seated structures of racial capitalism. This is not limited to institutions that are strictly environmental or climate focused. Carceral systems, for example, are fundamental to processes that create differential value that make racial capitalism function (Gilmore 2007; Pulido 2017). Increasing attention by activists and scholarship by academics connects police violence, school-to-prison pipelines, and other elements of carceral systems to environmental racism and injustice (Dillon and Sze 2016; Williams 2021). Criminalization of houselessness (Goodling 2020; Klein 2021), cruel and dehumanizing immigration enforcement, and ongoing abandonment of infrastructure systems (Pulido 2016; Ranganathan 2016; Silver 2019) are all products and reinforcements of the structures of racial capitalism. These systems manufacture group-differentiated vulnerability to premature death (Gilmore 2017). When speaking of vulnerabilities to climate hazards, lack of access to the benefits of decarbonization efforts, or the increasing segregation of the climate privileged and precarious (Rice et al. 2022), we are necessarily speaking of manufactured vulnerabilities and injustices. When speaking of abolitionist climate justice, then, we mean that we must take seriously creating a different world, free from police violence, void of cages, and reparative of the legacies of transatlantic slavery.

Critical environmental justice literatures provide insight into the practices needed to repair places that are made to let people die (Pellow 2017). The first step is to abolish institutions that perpetuate racialized violence in order to maintain and produce environments of the climate privileged. While we work toward abolition, a continual struggle, we must also engage in critical practices of repair and maintenance as a new politics of care.

Politics of Care

To create secure and equitable urban climate futures, we must work toward urban repair and maintenance in places that need it most. Urban climate justice cannot only be about building big and new solutions detached from place and community. Instead, we should focus on improving the lives of urban residents in ways that meet their everyday needs, wants, and desires. This involves recognition of historical oppression and present circumstances. It will take (re)building cities made for capital accumulation that have been accessible only for the privileged few. Repair and maintenance, here, is not synonymous with maintenance of oppressive social structures and processes. Instead, it is more literal. We must repair failing infrastructures, mend communities ripped apart by oppressive property relations and punitive systems, and care for each other and our life-supporting systems under new regimes of cooperative, community-driven management. This will be challenging, but motivated by an abolitionist climate politics, the struggle can be productive of something altogether new and better.

Recent scholarship has highlighted the extent to which racialized cities are in disrepair, with vital infrastructures that fail and further entrench inequalities (Pulido 2016; Ranganathan 2016; Silver 2019). Beyond the context of the United States, this is often discussed as a split between "formality" and "informality," between the networked ideal of infrastructural modernity and the fractured reality of piecemeal infrastructural provision. However, as Baptista (2019) argues, "the relentlessness of infrastructure maintenance and repair is more prominent in contexts of 'informality,' where on-the-ground conditions are challenging and resources scant" (516). To be sure, the ongoing challenges of infrastructural maintenance and repair are more pronounced where there is exclusion, whether that be in forms of neocolonialism or environmental racism.

Yet increased scholarly attention to the way infrastructures have always been incremental and improvisational has opened windows into possibilities for future configurations that center equity and justice. Indeed, what led many scholars to focus so heavily on urban climate "experiments" (Levenda 2019) was the political possibilities that trialing, testing, and learning have for urban climate justice. This is not a valorization of "innovation" but a push to imagine cities as venues for harnessing the power of community and collectivity. Repair and maintenance can center democratic commons, revalue public infrastructures, and create low-carbon, climate-safe futures.

More generally, maintenance and repair represent a politics of care (Mattern 2018). This includes care for physical infrastructures through objects like pipes, wires, streets, trains, buildings, and parks, but it also includes care for social infrastructures and our relationships with one another. Feminist perspectives on the politics of care point toward the need to dismantle oppressive systems upheld in heteropatriarchy (Arvin, Tuck, and Morrill 2013; Martin, Myers, and Viseu 2015). Social reproduction, broadly, is a key component of understating care as socially necessary labor, that is, the very maintenance of human life. As Millington (2019) notes, "Repair can also be a care practice, especially if we understand the infrastructures that surround us to be interlinked in complex, intimate ways with broader dynamics of social reproduction." But care also is much more than reproductive capacities. It can be described as "everything we do to maintain, continue, and repair 'our world' so that we can live in it as well as possible" (Fisher and Tronto 1990, 40). Care, then, can be generalized as politics in itself that entails maintenance and repair of things, infrastructures, relationships, people, ecosystems, communities, and institutions.

Care takes on a new urgency in the context of impending climate chaos (Klein 2020). As numerous scientific reports and, indeed, warnings of increasing chances of runaway climate catastrophe emerge year after year, the general sense of the need for action has heightened. These calls for action come not only from political policy makers at all levels of government but more importantly from frontline, vulnerable communities who are calling for socioecological transformations that equate action with a right to live, a right to stay put, and a right to a future. These calls from the most climate precarious are necessary to recognize and act on. This form of recognition can be seen as a form of care, an ethico-political commitment to climate justice by and for those who are most at risk. For example, what would it mean to reject private landownership in favor of community land trusts and massive amounts of green public housing? How can we enact initiatives like those proposed by Bernie Sanders and Alexandria Ocasio-Cortez's Green New Deal for Public Housing Act, which seeks to "retrofit, rehabilitate, and decarbonize the entire nation's public housing stock"? (Sanders 2019). What steps can begin to undo the criminalization of climate migrants, where the construction of a border wall often does far more to harm the environment than any group of people seeking refuge? In the mainstream political menu of outright antiscience climate denialism and climate denialism via incrementalism, caring involves providing alternative paths and acting out visions of radical and abolitionist climate justice. Urban climate justice must entail care for one another and our life-support systems. As Indigenous scholar Robin Wall Kimmerer (2013) explains, "all flourishing is mutual" (15).

Reparations

How might we get to this imagined place of just urban climate futures? If we take seriously calls for abolitionist climate justice and the politics of care this requires, we must also take climate reparations seriously. Scholars and activists have increasingly called for greater attention to mechanisms that account for globally unequal exchange, ecological debt, and uneven contributions of climate change–causing greenhouse gases. While multiple proposals have been outlined, the only tenable solutions for addressing past harms and beginning to heal trauma of root causes is a form of reparations that acknowledges the interrelated socioecological processes of oppression that have created the climate crisis. Maxine Burkett (2009) outlines the basic tenets of climate reparations—a form of identification, acknowledgment, and compensation for past wrongs seeking to improve lives moving forward. Burkett explains:

> Climate reparations is the effort to assess the harm caused by the past emissions of the major polluters and to improve the lives of the climate vulnerable through direct programs, policies, and/or mechanisms for significant resource transfers, to assure the ability of the climate vulnerable to contemplate a better livelihood in light of future climate challenges. In order to repair individual communities, as well as the global community, all those engaged in the reparative effort will have to squarely confront the deep moral questions posed by both the initiating harm—excess emissions—and the continuing harm: the failure to adequately include the plight of the climate vulnerable in the current processes developed to mitigate and adapt to the climate crisis. (523)

In Burkett's formulation, the climate precarious are centered and redistribution is required. The practicalities of how reparations function, what form they take, and for whom are of central concern in Burkett's discussion, always with a direction toward community control and wrongdoers' (i.e. colonial powers, oppressive domineers of capitalist imperialism) responsibility for action on mitigation.

Likewise, Moulton and Machado (2019) argue that, in the context of Caribbean islands, "reparations for the developmental and climate debt owed to the region offer both ethical and practical paths to redressing the consequences of two and a half centuries of industrialization and overdevelopment in the 'First World' and the unequal distribution of the wealth that is tied to such processes" (16). Similarly, Perry (2020) critiques the sustainable development goals' (SDGs) lack of consideration of climate reparations and role in furthering socioecological harm, and holds reparations as essential: "Remedying damages and losses through a reparatory agenda are at the centre of climate justice for those made vulnerable through imperial designs, deliberate historical acts, disregard and even industrial violence" (19). Within the context of international law and political economy scholarship, the agenda for climate reparations has been discussed and is receiving more attention; yet the issues are still marginalized by technocratic agendas pushed by the footholds of U.S. and European imperial capitalist power.

Sheller (2020) also argues that reparations are essential for a future marked by increasing climate-driven displacements and mobilities. Sheller explains, from a U.S. positionality, that regardless of the form of climate reparations, "we must reject the depiction of climate refugees as a growing danger who will 'flood' our borders." Indeed, numerous calls for mobilities justice in the form of immigration and permanent resettlement, or even strategic and managed retreat, will increasingly mark the contours of urban climate justice moving ahead. Furthermore, given current tendencies toward rising ethnonationalism, eco-fascism, and anti-migration sentiment in many countries, this is a crucial area for future research and activism in relation to climate change. Beyond the potential for segregation of cities and securitization of borders in the Global North, there are elsewhere forms of ethnonationalism, casteism, and other racializations that stand to be exacerbated by the climate crisis, and these must be researched and addressed beyond Western-liberal contexts.

Ultimately, how well we create inclusive and life-affirming systems for integration and relocation will be a testament to how we center climate justice. Examples such as sanctuary cities movements and home buyback programs paint a rather bleak picture. Instead, we need visionary and radical changes that locate power in people, ensure freedom, and center liberation. This means reparations cannot simply be a metaphor. Rather, climate reparations require new ways to transfer and redistribute wealth, resources, land, and property that have been held by an elite few in the climate-changed city.

Within the context of marginalized and disinvested cities of the North, the role of reparations is gaining salience. Frontline communities in places such as the United States Gulf South have pushed for climate action that engages reparative efforts for the socioecological plunder of petrochemical industries and climate disasters. Here, major industrial actors as well as local, state, and federal governments are complicit in upholding the continued marginalization of the climate precarious. Accountability, community-driven projects, and reparations are a central thread of climate justice agendas, such as the Gulf South for a Green New Deal (Carmichael 2020) and the Red, Black, and Green New Deal (www.redblackgreennewdeal.org). Movements such as these are inspiring for transformative climate justice built on tenets of abolition, care, and reparations. These are the horizons of urban climate justice.

Conclusion: The Future

We, and the authors who contributed to this volume, maintain a forward-looking, hopeful, and transformative view of the climate-just city. While climate change has exacerbated environmental vulnerability and hazards in urban areas throughout the world, the chapters in this book give readers a clear understanding of the root causes of inequity and injustice and provide a set of actions and approaches that move us closer to the horizon of climate justice. We envision the overarching lessons and ideas in this text to be used by scholars and practitioners in their own everyday practice, but with an eye toward a planetary, inclusive, and democratic perspective.

The majority of the contributions in this book have come from scholars based in the United States and United Kingdom, and as a result, we recognized that we must first critique an unjust system whose influence still largely emanates from such historic seats of colonial power. As of the writing of this book, large-scale climate policies and actions are embedded within an international political economy dominated by the most influential nation-states, global cities, financial institutions, and development organizations of the Global North. Because they are built on and permeated by the persistent legacies of colonialism, heteropatriarchy, and racial capitalism, any climate policies and actions that originate within this system risk reinforcing and perpetuating injustice. A major motivation for this book has been to confront these legacies head-on to stop the cyclical production of privilege and precarity. Doing so requires us to interrogate and disrupt our own positionalities, but also to recognize that our work cannot know or understand all the ways that injustice manifests.

With that in mind, we hope that readers will engage these chapters critically and practically, situating each contribution within a broader discourse of justice. Doing so may remind us to engage in both internal and external critiques of the structures and regimes that perpetuate the climate crisis. It may also encourage us to strengthen and participate in local to global networks of dialogue, resistance, and intervention. Ultimately, we recognize that urban climate justice is a collective action—one that requires purposeful activism, conscientious solidarity, and unbounded allyship in the pursuit of existential planetary justice.

REFERENCES

Arvin, M., Tuck, E., and Morrill, A. (2013). Decolonizing feminism: Challenging connections between settler colonialism and heteropatriarchy. *Feminist Formations*, 25(1), 8–34.

Baptista, I. (2019). Electricity services always in the making: Informality and the work of infrastructure maintenance and repair in an African city. *Urban Studies*, 56(3), 510–525.

Burkett, M. (2009). Climate reparations. *Melbourne Journal of International Law*, 10, 509.

Carmichael, E. (2020). Gulf South communities say climate justice requires new economic systems. *Scalawag*, February 11. https://scalawagmagazine.org/2020/02/green-new-deal-gulf-coast/.

Dawson, A. (2010). Climate justice: The emerging movement against green capitalism. *South Atlantic Quarterly*, 109(2), 313–338.

Dillon, L., and Sze, J. (2016). Police power and particulate matters: Environmental justice and the spatialities of in/securities in U.S. cities. *English Language Notes*, 54(2), 13–23.

Fisher, B. and Tronto, J. C. (1990). Toward a feminist theory of care. In E. K. Abel and M. K. Nelson (Eds.), *Circles of Care: Work and Identity in Women's Lives (pp. 35–62)*. SUNY Press.

Gilmore, R. W. (2007). *Golden Gulag: Prisons, Surplus, Crisis, and Opposition in Globalizing California*. University of California Press.

———. (2017) Abolition geography and the problem of innocence. In G. T. Johnson and A. Lubin (Eds.), *Futures of Black Radicalism* (pp. 225–240). Verso.

Goodling, E. (2020). Intersecting hazards, intersectional identities: A baseline Critical Environmental Justice analysis of U.S. homelessness. *Environment and Planning E: Nature and Space*, 3(3), 833–856.

Kimmerer, R. W. (2013). *Braiding Sweetgrass: Indigenous Wisdom, Scientific Knowledge and the Teachings of Plants*. Milkweed.

Klein, N. (2020). Care and repair: Left politics in the age of climate change. *Dissent*, Winter. https://www.dissentmagazine.org/article/care-and-repair-left-politics-in-the-age-of-climate-change.

———. (2021). A climate dystopia in Northern California. *Intercept*, May 7. https://theintercept.com/2021/05/07/california-fires-chico-housing-real-estate/.

Levenda, A. M. (2019). Mobilizing smart grid experiments: Policy mobilities and urban energy governance. *Environment and Planning C: Politics and Space*, 37(4), 634–651.

Martin, A., Myers, N., and Viseu, A. (2015). The politics of care in technoscience. *Social Studies of Science*, 45(5), 625–641.

Mattern, S. (2018). Maintenance and care. *Places*. https://placesjournal.org/article/maintenance-and-care/.

Millington, N. (2019). Critical spatial practices of repair. *Society and Space*. https://www.societyandspace.org/articles/critical-spatial-practices-of-repair.

Moulton, A. A., and Machado, M. R. (2019). Bouncing forward after Irma and Maria: Acknowledging colonialism, problematizing resilience, and thinking climate justice. *Journal of Extreme Events*, 6(1).

Pellow, D. (2017). *What Is Critical Environmental Justice?*. John Wiley & Sons.

Perry, K. (2020). The New "Bond-Age," Climate Crisis and the Case for Climate Reparations: Unpicking Old/New Colonialities of Finance for Development within the SDGs (SSRN Scholarly Paper ID 3739103). *Social Science Research Network*. https://doi.org/10.2139/ssrn.3739103.

Pulido, L. (2016). Flint, environmental racism, and racial capitalism. *Capitalism Nature Socialism*, 27(3), 1–16.

——. (2017). Geographies of race and ethnicity II: Environmental racism, racial capitalism, and state-sanctioned violence. *Progress in Human Geography*, 41(4), 524–533.

Ranganathan, M. (2016). Thinking with Flint: Racial liberalism and the roots of an American water tragedy. *Capitalism Nature Socialism*, 27(3), 17–33.

Ranganathan, M., and Bratman, E. (2021). From urban resilience to abolitionist climate justice in Washington, D.C. *Antipode*, 53(1), 115–137.

Rice, J., Long, J., and Levenda, A. (2022). Against climate apartheid: Confronting the persistent legacies of expendability for climate justice. *Environment and Planning E: Nature and Space*, 5(2), 625–645.

Sanders, B. (2019). Sanders and Ocasio-Cortez Announce the Green New Deal for Public Housing Act. Bernie Sanders U.S. Senator for Vermont, November 14. https://www.sanders.senate.gov/press-releases/sanders-and-ocasio-cortez-announce-the-green-new-deal-for-public-housing-act/.

Sheller, M. (2020). The case for climate reparations. *Bulletin of the Atomic Scientists*, November 6. https://thebulletin.org/2020/11/the-case-for-climate-reparations/.

Silver, J. (2019). Decaying infrastructures in the post-industrial city: An urban political ecology of the U.S. pipeline crisis. *Environment and Planning E: Nature and Space*. https://doi.org/10.1177/2514848619890513.

Williams, T. (2021). For "peace, quiet, and respect": Race, policing, and land grabbing on Chicago's South Side. *Antipode*, 53(2), 497–523.

CONTRIBUTORS

Dietrich Bouma is a doctoral student in city and regional planning at Cornell University. His research investigates equitable and just approaches to plan and govern climate-induced (im)mobilities for marginalized communities.

Vanesa Castán Broto is a professor of climate urbanism at the University of Sheffield. Her current research on alternative and mundane urban innovation for climate change is funded by the European Research Council and the UK's Global Challenges Research Fund. Her books include *Urban Energy Landscapes* (Cambridge University Press, 2019) and *Urban Sustainability and Justice: Just Sustainabilities and Environmental Planning* (ZED Books, 2002) and the coedited collection *Climate Urbanism: Towards a Critical Research Agenda* (Palgrave, 2021).

Savannah Cox is a PhD candidate at UC Berkeley's Department of City and Regional Planning. Her work addresses the intersections of climate finance and climate justice with a topical focus on resilient infrastructure investment in cities.

Joan Fitzgerald is a professor of urban and public policy at Northeastern University. She focuses on urban climate action and strategies for linking it to equity, economic development, and innovation. In her fourth book, *Greenovation: Urban Leadership on Climate Change* (Oxford University Press, 2020), she discusses cities leading the way on climate change strategies in North America and Europe and offers strategies for cities lagging in such innovations to accelerate their action.

Sheila R. Foster is the Scott K. Ginsburg Professor of Urban Law and Policy at Georgetown University. She holds a joint appointment with the Georgetown Law School and the McCourt Public Policy School. Foster writes in the areas of property, land use, environmental justice, and local government law. She is well known for her articles and books on environmental justice, including *From the Ground Up: Environmental Racism and the Rise of the Environmental Justice Movement* (with Luke Cole; NYU Press, 2000) and *The Law of Environmental Justice* (with Michael Gerrard; American Bar Association, 2009). She is a member of the New York City Mayor's Panel on Climate Change and cochairs the panel's workgroup on equity.

Kian Goh is an associate professor of urban planning at the University of California, Los Angeles, and an associate faculty director of the UCLA Luskin Institute on Inequality and Democracy. She researches urban ecological design, spatial politics, and social mobilization in the context of climate change and global urbanization. She is the author of *Form and Flow: The Spatial Politics of Urban Resilience and Climate Justice* (MIT Press, 2021).

Ping Huang is a postdoctoral researcher at the Urban Institute of the Interdisciplinary Centre of Social Sciences at the University of Sheffield. He is also a visiting scholar at the Institute for International Affairs, Qianhai, at the Chinese University of Hong Kong, Shenzhen. His primary research interests are green and low-carbon innovation management and policy, urban sustainability transitions, and climate governance.

Sarah Knuth is an assistant professor in the Department of Geography at Durham University, United Kingdom. Her research critically investigates how global climate change and responses to it intersect with preexisting drivers of precarity, inequality, and injustice. She focuses on how new forms of financial speculation, extraction, and self-protection are unfolding within fast-changing energy systems and urban economies and with what implications for frontline communities. Another of her priorities is identifying openings for more just climate futures and pathways to self-determination.

Ankit Kumar is a lecturer in development and environment at the Department of Geography, the University of Sheffield. Ankit's research interests are situated around climate and energy justice in the Global South and driven by critical development studies, postcolonial studies, and environmental geographies. He has published in several journals, including *Transactions of the Institute of British Geographers, Social and Cultural Geography, Antipode,* and *Energy Research and Social Science.*

Robin Leichenko is a professor of geography at Rutgers University and the codirector of the Rutgers Climate Institute. Her current research explores the social, economic, and equity implications of climate change with a focus on the northeastern United States. Her latest book, *Climate and Society: Transforming the Future* (coauthored with Karen O'Brien), was published by Polity Press in 2019.

Anthony Levenda is the director of the Center for Climate Action and Sustainability at the Evergreen State College. His work focuses on the politics of urban sustainability, decarbonization, and resilience.

Joshua Long is an interdisciplinary human geographer whose research focuses on urban theory, environmental justice, environmental politics, and spatial theory. In recent years, he has focused his attention on climate justice and the complexity of addressing the climate crisis equitably, ethically, and inclusively. Long is currently a professor of environmental studies and chair of the Environmental Studies Program at Southwestern University.

Diego Martinez-Lugo holds an MA in geography and an MS in Mexican American studies from the University of Arizona (2020). Diego's research uses ethnic studies and geography to examine how race, power, class, and space entangle to coproduce environmental and climate justice.

Khai Hoan Nguyen is a PhD candidate in the Department of Geography at Rutgers University. Her research focuses on the relationship between land use, economic development, and urban climate adaptation with specific interests in planning and policy, industry, and social equity.

Jenny Pickerill is a professor of environmental geography and head of the Department of Geography at Sheffield University, England. Her research focuses on grassroots solutions to environmental problems and hopeful and positive ways in which we can change social practices. She has published three books—*Cyberprotest: Environmental Activism Online* (Manchester University Press, 2003), *Anti-war Activism: New Media and Protest in the Information Age* (Palgrave, 2008), and *Eco-homes: People, Place, and Politics* (Bloomsbury, 2016)—and more than thirty articles on themes around eco-housing, eco-communities, social justice, and environmentalism. She is currently completing a book titled *Eco-communities: Living Together Differently.*

Vanessa Raditz (they/them) is an educator, researcher, and cultural organizer dedicated to community healing, opening access to land and resources, and fostering a thriving local economy based on human and ecological resilience. Raditz received a master of public health in environmental health sciences from UC Berkeley, and is currently a graduate student in the Department of Geography at University of Georgia, where their research uses digital storytelling to explore the cultural geographies of queer ecojustice movements. Raditz is part of the founding collective of the Queer Ecojustice Project, educating and organizing at the intersection of ecological justice and queer liberation.

Jennifer L. Rice is an associate professor in the Department of Geography at the University of Georgia. She is an urban geographer interested in concrete strategies for social and environmental justice. Her longest-running project focuses on urban climate justice and resisting climate apartheid. She also examines housing justice in the "tech city" and racial justice in the wake of settler-colonial urbanization and higher education.

Enora Robin is a Leverhulme Postdoctoral Research Associate at the Urban Institute, University of Sheffield. Her research focuses on critical analyses of financing mechanisms that are deployed to facilitate transitions to low-carbon cities. She is coeditor of the collection *Climate Urbanism: Towards a Critical Research Agenda* (Palgrave, 2021).

Gloria Schmitz is a doctoral candidate at Northeastern University's School of Public Policy and Urban Affairs in Boston, Massachusetts. Her program concentration is sustainability and resilience, and her dissertation research explores how the

COVID-19 pandemic has affected the transition to circular economies both in Boston and worldwide. Her hobbies include running, yoga, and traveling the world to experience the richness of other cultures.

Linda Shi is an assistant professor at Cornell University's Department of City and Regional Planning. She studies how urban land governance shapes climate vulnerability and the equity impacts of climate adaptation responses.

Jonathan Silver is a senior research fellow based at the Urban Institute, University of Sheffield. His work focuses on thinking through the geographies of infrastructure as they are planned, operated, and experienced.

Jennie C. Stephens is the Dean's Professor of Sustainability Science and Policy at Northeastern University's School of Public Policy and Urban Affairs. She is an internationally recognized expert on renewable energy transformation, energy justice, climate justice, energy democracy, and gender and race in energy and climate. Her most recent book, *Diversifying Power: Why We Need Antiracist, Feminist Leadership* (Island Press, 2020), inspires collective action by elevating the stories of innovative, diverse leaders who are linking climate and energy with jobs and economic justice, health and food, and transportation and housing.

Linda Westman is a research associate at the Urban Institute, University of Sheffield. Her work engages with the governance of climate change, cities, transformation, and justice. She is the chapter scientist of and a contributing author to WGII, chapter 6 of the Intergovernmental Panel on Climate Change, and a chapter author for the UN-Habitat 2022 World Cities Report.

INDEX

abolition, 20, 68, 78, 79, 202, 231, 253, 254–256. *See also* abolitionist

abolitionist, 12, 79, 199; as form of climate justice, 11, 12, 193, 255, 256, 257

access: to climate security, 1, 14, 90; to finance, 90, 93, 96, 244; to food, 116, 139, 142; functional needs and, 131, 133, 134, 136; to housing, 114, 117–118, 120, 247; to land, 10, 34, 48; to renewable energy, 110, 118, 123n2; to resilient infrastructure, 1, 90; to resources, 16, 34, 38, 114; to services or amenities, 6, 34, 80, 138, 148, 244, 248; to technology, 110; to transportation, 116, 117, 121, 122

accessible, 4, 14, 33, 70, 240, 257

anti-Black/Blackness, 10, 178, 191, 245

anticolonial, 17, 18, 21, 89, 90, 91, 102, 222, 228, 229, 231, 253. *See also* colonization

assemblages, 174, 208

Audre Lorde Project, 167

Austin, Tex., 122

authoritarian, 153

biodiversity, 175, 247

BIPOC (Black and Indigenous People of Color), 3, 19, 177–178, 246, 247; in QTBIPOC, 167–168, 170–173, 176, 180, 181nn1–2

BLM (Black Lives Matter), 17, 40, 60, 68, 72, 78–79, 82, 138, 195, 221

borders, 3, 48, 62, 191, 230, 260

Boston, Mass., 19, 109, 112–124

Boston Housing Authority, 115

carbon: capitalism, 10, 70, 90, 91, 95–96, 101, 191; colonialism, 3; in counting or budgets, 1, 39; debt, 39; fetishization, 7; financial instruments, 74; gentrification, 3; governmentality, 5; pricing or markets, 69, 95–96, 117; taxation, 32. *See also* decarbonize

Carbon Free Social Equity Report, 114, 115, 116, 120

Carbon Neutral Cities Alliance (CNCA), 112, 113

Ciliwung Merdeka, 154, 156–157

Ciliwung River, 156, 157, 160

cisgender. *See* gender

Clean Development Mechanism (CDM), 90, 94, 96–97, 101

climate action plans, 112, 114–120

climate apartheid, 12, 97, 223, 255

climate gentrification, 3, 14, 16, 52, 57, 141, 150, 215, 247

climate urbanism, 5, 7, 12, 13, 47, 89–91, 93–102, 150, 215, 216, 236, 239

collectivity, 223, 236–237, 242, 244, 257

colonization, 2, 5, 8, 9, 13, 54, 179, 190, 191, 222, 229

contextual equity. *See* equity

coronavirus, 1. *See also* COVID-19

corporations, 1, 55; climate change work of, 14; energy and, 142; equity and, 128, 130; finance and, 97; landownership of, 50

counterplanning, 156

COVID-19, 1, 40, 121, 122, 127, 176, 202

Dakota Access Pipeline (DAPL), 9, 192

decarbonization, 17, 67, 81, 256, 258

decolonization, 185, 201, 254

disability, 136, 174

disabled, 169, 172, 174, 240

GEOGRAPHIES OF JUSTICE AND SOCIAL TRANSFORMATION